The ABC of Acid-Base Chemistry

The ABC of Acid-Base Chemistry

The Elements of Physiological Blood-Gas Chemistry
for Medical Students and Physicians

Horace W. Davenport

Sixth Edition, Revised

The University of Chicago Press
Chicago & London

The University of Chicago Press, Chicago 60637
The University of Chicago Press, Ltd., London

© 1947, 1949, 1950, 1958, 1969, 1974 by The University
of Chicago

International Standard Book Number:
0–226–13705–8 (clothbound); 0–226–13703–1 (paperbound)
Library of Congress Catalog Card Number: 73–90943

Contents

Preface

The purpose of this book is to provide medical students and physicians with an intelligible outline of the elements of physiological acid-base chemistry. The subject is of such great practical importance that its serious study is almost obligatory. Unfortunately, there are no effortless roads to a knowledge of acid-base chemistry, and there are no easily memorized rules-of-thumb which can be applied at all times in the clinic without risk of disaster. Consequently, a student wishing to master the subject must work earnestly and seriously, and I hope this book will serve as a reliable guide in his initial studies. Once he has mastered the principles, the more difficult aspects of the subject encountered in research or at the bedside are easy to grasp.

The well-trained biochemist or physician who reads this book will discover that it contains many elisions. In places I have tailored the exposition to fit my judgment of what is both sensible and important. At many points I have decided that the advantages of simplicity and clarity in an elementary text outweigh the advantages and elegance of strict accuracy and thermodynamic logic. Finally, I leave the application of the subject to the students' professors of medicine, surgery, and pediatrics.

I am indebted to many persons, particularly to Professor A. Baird Hastings and the late Robert Wolf. I am grateful to the late Professor Franklin C. McLean, to the late Virginia D. Davenport, R. E. Forster, II, and to the many others who have given me help. I am especially grateful to Arthur J. Vander for his counsel throughout the preparation of the last two editions. I must especially thank the hundreds of students who, over the past twenty-five years, have pointed out errors or ambiguities.

The ABC of Acid-Base Chemistry

What Happens in Blood 1

1.1. The Partial Pressure of a Gas

The chemical and physiological actions of a gas depend upon the pressure it exerts. The pressure which any one gas exerts, whether it is alone or mixed with other gases, is the *partial pressure* of the gas. The partial pressure of a gas is usually denoted by the small letter p preceding the symbol for the gas; thus pO_2 represents the partial pressure of oxygen. Respiratory physiologists denote a partial pressure by the capital letter P followed by the symbol as a subscript; thus P_{O_2} is for them the partial pressure of oxygen. Since this book will be used by persons studying respiratory physiology, this system of symbols will be used.

The partial pressure of a gas depends only on the number of moles of gas in a given volume and on the temperature. It is independent of the presence of other gases in the same volume. This fact is represented by the perfect gas equation which, when applied to oxygen is

$$P_{O_2} = N_{O_2} RT/V. \tag{1}$$

In this equation N_{O_2} represents the number of moles of oxygen in the volume V; the quantity R is the gas constant which is the same for all gases; and T is the absolute temperature.

The total pressure exerted by a gas mixture such as atmospheric air is the simple arithmetical sum of the partial pressures of the gases making up the mixture. Barometric pressure (P_B) is therefore the sum of the partial pressures of oxygen, carbon dioxide, and nitrogen of the air. This fact is represented by the equation

$$P_B = P_{CO_2} + P_{O_2} + P_{N_2}. \tag{2}$$

Since the partial pressures of the component gases in a mixture are proportional to the number of moles of gas present, the equations

$$P_{CO_2} = N_{CO_2} RT/V \tag{3}$$

and

$$P_{N_2} = N_{N_2} RT/V \tag{4}$$

can be written. Barometric pressure is equal to the sum of equations (1), (3), and (4):

$$P_B = (N_{O_2} + N_{CO_2} + N_{N_2})RT/V. \tag{5}$$

Dividing equation (1) by equation (5), we have

$$\frac{P_{O_2}}{P_B} = \frac{N_{O_2}}{N_{O_2} + N_{CO_2} + N_{N_2}}. \tag{6}$$

This equation states that the ratio of the partial pressure of oxygen to the barometric pressure is equal to the ratio of the number of moles of oxygen to the total number of moles of gas. The right hand term in the equation represents the fraction of oxygen in the gas mixture. The equation can be written as

$$P_{O_2} = P_B(\text{fraction of } O_2 \text{ moles in air}) = P_B(F_{O_2}). \tag{7}$$

Equal volumes of gas contain equal numbers of moles, and consequently the fraction of gas by volume is also equal to the fraction of gas by moles. The fraction of a gas by volume is the percentage divided by 100. Thus the fraction of oxygen in a gas mixture equals $\%O_2/100$. When this is substituted in equation (7) the equation becomes

$$P_{O_2} = P_B(\%O_2)/100. \tag{8}$$

Equations for the partial pressures of other gases can be written

$$P_{CO_2} = P_B(\%CO_2)/100, \tag{9}$$

and

$$P_{N_2} = P_B(\%N_2)/100. \tag{10}$$

Repeated analysis of air has shown that its percentage composition is

$$\%O_2 = 20.93,$$
$$\%CO_2 = 0.03,$$

and

$$\%N_2 = 79.04.$$

If the barometric pressure is known, the partial pressure of the gases can be calculated.

Example 1. If the barometric pressure of dry air is 760 mm Hg, what are the partial pressures of oxygen, carbon dioxide, and nitrogen?

$$P_{O_2} = (760 \text{ mm Hg})(20.93)/100 = 159 \text{ mm Hg},$$
$$P_{CO_2} = (760 \text{ mm Hg})(0.03)/100 = 0.2 \text{ mm Hg},$$

and

$$P_{N_2} = (760 \text{ mm Hg})(79.04)/100 = 601 \text{ mm Hg}.$$

The partial pressure of a gas may be determined by chemical or physical means. When chemical means are used, percentage composition of the gas by volume is measured; and partial pressures are calculated from the percentages and the barometric pressure. Individual gas species have physical properties upon which measurement of their partial pressure is based. Carbon dioxide absorbs infrared radiation, and a carbon dioxide meter which measures the degree of absorption of a beam of infrared radiation passing through the gas sample gives a meter reading which is a function of the carbon dioxide partial pressure in the sample. Consequently, the meter can be calibrated with samples of carbon dioxide having a known pressure, and the meter reading can be converted into partial pressure reading. In a similar manner, meters which measure the partial pressure of oxygen by means of its paramagnetic property or of nitrogen by means of the light it emits in an electric field under high vacuum have been constructed. When the partial pressure is known and it is desired to calculate the percentage or fractional composition of the gas, equations similar to (7) are used; and the known partial pressure and barometric pressure are substituted in them, thus allowing the unknown fraction to be calculated.

1.2. Composition of Alveolar Air

Air breathed in during inspiration—the inspired air—mixes with gas already present in the trachea, the bronchi, and the finer subdivisions of the lungs. Some of this mixture is drawn into the expanding alveoli where it comes into contact with the lung capillaries carrying venous blood. This gas mixture in the alveolar spaces is called *alveolar air*. From it some oxygen passes into the blood, and into it some carbon dioxide is released by the blood. Upon expiration, some of the alveolar air is forced out of the alveoli into the bronchi and trachea where it mixes with gas present in these passages, and the mixture is breathed out as expired air.

Under steady-state conditions tidal flow of gas into and out of alveolar spaces maintains constancy of the composition of alveolar air. On the average, just as much oxygen comes into the alveolar air from inspired air as is removed from alveolar air by the blood, and as much carbon dioxide is expired as is delivered into alveolar air by the blood.

Alveolar air contains the three gases of inspired air— oxygen, carbon dioxide, and nitrogen; but it is also saturated with water vapor which evaporates from the surfaces of the tissues. Water vapor is a gas which exerts a partial pressure in the same manner as any other gas. However, the partial pressure of water vapor in a mixture which is saturated with it is a function of the temperature only. It is independent of the pressure of the other gases and of the barometric pressure. At normal body temperature of 37°C the partial pressure of water vapor is 47 mm Hg.

The total pressure of gases in alveolar air is the same as barometric pressure. Since 47 mm Hg of that pressure is necessarily taken up by water vapor, the pressure of the other gases, the dry gases, is the barometric pressure minus 47 mm Hg. When alveolar air is collected and analysed by chemical means, its composition is reported in terms of percentage of oxygen and carbon dioxide in dry gas, or in gas minus water vapor. The actual partial pressure of one of the gases in alveolar air is therefore calculated by multiplying the fraction of gas in dry air by the barometric pressure minus 47 mm Hg.

> Example 2.   Barometric pressure is 745 mm Hg. A sample of alveolar air is collected and analysed. Carbon dioxide and oxygen percentages are 5.6 and 14.8, respectively. What are the partial pressures of the gases in alveolar air?
>
> $$P_{CO_2} = (745 - 47 \text{ mm Hg})(5.6)/100 \ = \ 39.1 \text{ mm Hg},$$
> $$P_{O_2} = (745 - 47 \text{ mm Hg})(14.8)/100 = 103.3 \text{ mm Hg}.$$

In normal man at rest and breathing quietly, the respiratory mechanism maintains the P_{CO_2} of alveolar air at about 40 mm Hg. As the metabolic production of carbon dioxide rises from the resting value of about 250 cc a minute to about 2,500 cc a minute attained in heavy exercise, alveolar ventilation rises in exact proportion. Consequently, over this wide range of carbon dioxide production the P_{CO_2} of alveolar air remains constant at nearly 40 mm Hg. Constancy of alveolar P_{CO_2} is disturbed when alveolar ventilation increases or decreases out of proportion to the rate of carbon dioxide production. Hyperventilation occurs in man at rest when there is an urgent demand for oxygen, as in high-altitude hypoxia; when there are reflex or central nervous stimuli increasing ventilation as in pain, high body temperature, hysteria, or salicylate intoxication: or when there is a demand for respiratory compensation for metabolic acidosis. Then alveolar P_{CO_2} falls below 40 mm Hg. Hypoventilation occurs when the respiratory apparatus is unable for any reason to provide adequate alveolar ventilation; then alveolar P_{CO_2} rises above 40 mm Hg.

In a man at rest not suffering from hypoxia, the P_{O_2} of alveolar air is not regulated with the same degree of precision as is the P_{CO_2}. The P_{O_2} of alveolar air in a normal man at sea level is about 100 mm Hg. Hyperventilation raises the alveolar P_{O_2}; and when room air is breathed, the upper limit of alveolar P_{O_2} is about 140 mm Hg. Hypoventilation lowers alveolar P_{O_2}.

1.3. Carriage of Oxygen in the Blood

When the partial pressure of oxygen is different in two parts of a system, a diffusion gradient exists; and the gas diffuses from the place where its partial pressure is high to that where it is low. If the system is left undisturbed, the partial pressure of the gas becomes the same in all parts of the system. Then a diffusion gradient no longer exists, and the system is in equilibrium with respect to the partial pressure of the gas. The rate at which the gas diffuses depends in part on the steepness of the diffusion gradient, and the gas diffuses faster, the greater the difference in partial pressure between the two parts of the system. The rate at which the gas diffuses also depends upon the nature of the barrier between the parts of the system. If there is no barrier, the gas diffuses rapidly, and equilibrium is quickly reached. In the lungs, the alveolar membrane separating alveolar air from blood is a barrier to diffusion of oxygen, because oxygen is relatively poorly soluble in the water which forms the aqueous phase of the membrane.

Oxygen diffuses from alveolar air into blood, because venous blood flowing through the lungs has, except in unusual experimental circumstances, a P_{O_2} lower than that of alveolar air. Diffusion of oxygen into venous blood converts it into arterial blood. Blood flows rapidly through the lungs; when a man is at rest a single erythrocyte passes through lung capillaries in about 0.75 second; and when the man is exercising, the time is only 0.3 second. Because the alveolar membrane is a barrier to diffusion of oxygen, blood passing through the lungs never quite comes into equilibrium with alveolar air, and the P_{O_2} of blood leaving the capillaries of any particular alveolus is always slightly lower than the P_{O_2} of the air in the alveolus. In normal circumstances this difference is probably not greater than 1 mm Hg. However, arterial blood in systemic arteries of a man at rest has a P_{O_2} about 5 to 10 mm Hg lower than the P_{O_2} of alveolar air. Anatomical shunting of venous blood from the right ventricle to the left ventricle accounts for part of this difference, and regional variations in the relation of ventilation of alveoli to the perfusion of them by venous blood account for the rest. In severe exercise the difference between alveolar and arterial P_{O_2} increases to about 16 mm Hg, and in diseased states the difference may be even larger.

Oxygen contained in arterial blood is carried in two ways: as dissolved oxygen in physical solution and as a chemical compound with hemoglobin in the erythrocytes. The amount of oxygen dissolved and the amount combined with hemoglobin both depend on the P_{O_2} of arterial blood.

Arterial blood flows through the tissues where the P_{O_2} is lower than that of the blood. In some tissues, such as the brain and the thyroid gland, which have a very high rate of blood flow through them in relation to their oxygen consumption, the P_{O_2} may be only a few mm Hg lower than in arterial blood. In other tissues, such as exercising muscle, the P_{O_2} may be nearly zero. Because the P_{O_2} of the tissues is lower than that of arterial blood, oxygen diffuses from the blood into the tissues. The loss of oxygen from arterial blood, together with the concomitant gain in carbon dioxide, converts arterial blood to venous blood. Venous blood is collected in the veins, mixed in the right ventricle and again circulated through the lungs.

1.4. Carriage of Oxygen in the Blood by Physical Solution

The physical solution of oxygen in the blood follows a simple law which applies to all gases: the amount of oxygen dissolved in a given volume of blood is directly proportional to the P_{O_2} of the gas phase which is in equilibrium with the blood.

The law of solution of oxygen in blood is expressed by the equation

$$(O_2 \text{ dissolved}) = aP_{O_2}. \tag{11}$$

The constant a is the proportionality constant or solubility constant which can be measured. The solubility constant is often expressed in numerical terms such that it equals the number of cubic centimeters of gas dissolved in 1 milliliter of liquid when the partial pressure of the gas is 760 mm Hg or one atmosphere. Solubility constants, particularly for carbon dioxide as shown in section 1.13, are expressed in other units, and the student should be careful to note the units before using the constant.

The numerical value of a for oxygen in blood is 0.023 cc per milliliter of blood per atmosphere of oxygen at a temperature of 38°C. When the partial pressure of oxygen is less than one atmosphere, the amount of oxygen dissolved in 1 milliliter of blood is proportionately smaller. Accordingly, the amount of oxygen dissolved in 1 milliliter of blood is given by the equation

$$(O_2 \text{ dissolved}) = a \, P_{O_2}/760 \tag{12}$$
$$= (0.023) \, P_{O_2}/760 \text{ cc per ml.} \tag{13}$$

   The volume of oxygen dissolved in blood is conventionally expressed in volumes per cent, not in cc per ml. Volumes per cent, symbolized "vol %," is defined as the number of cubic centimeters of oxygen dissolved in 100 milliliters of blood. Consequently, the result as calculated by equation (13) must be multiplied by 100 to give the result in volumes per cent.

   Example 3.   Blood is equilibrated with a gas mixture containing 14.5% oxygen at a barometric pressure of 752 mm Hg and at 38°C. How many volumes per cent of oxygen are dissolved in the blood?

   First Step.   Calculate the P_{O_2}. Since the gas is in equilibrium with blood, it is saturated with water vapor. The total pressure of the dry gas is barometric pressure minus the pressure of water vapor:

$$\text{(Dry gas pressure)} = (752 - 47 \text{ mm Hg})$$
$$= 705 \text{ mm Hg.}$$

The P_{O_2} is the fraction of oxygen in the dry gas times the pressure of the dry gas:

$$P_{O_2} = \text{(Dry gas pressure)}(\%O_2)/100$$
$$= 705 (14.5)/100$$
$$= 102.2 \text{ mm Hg.}$$

Second Step.   The volume of oxygen dissolved in milliliter of blood is given by equation (13):

$$\text{(O}_2 \text{ dissolved)} = 0.023(102.2)/760$$
$$= 0.0031 \text{ cc per ml.}$$

Third Step.   Volumes dissolved in 100 milliliters is volumes per cent:

$$\text{(Vol \% dissolved)} = (0.0031 \text{ cc per ml)}(100)$$
$$= 0.31 \text{ vol \%.}$$

1.5. Carriage of Oxygen in Blood by Hemoglobin

Dissolved oxygen and hemoglobin react to form a chemical compound according to the equation

$$O_2 + Hb \;\rightleftharpoons\; HbO_2. \tag{14}$$

   The process by which hemoglobin is converted from oxygenated to deoxygenated hemoglobin is called *deoxygenation*. It is also frequently called *reduction*, although the iron in both oxygenated and deoxygenated hemoglobin is in the reduced or ferrous state. The process by which deoxygenated hemoglobin is converted to oxygenated hemoglobin is called *oxygenation*;

it is never called *oxidation.* Hemoglobin whose iron is oxidized to the ferric state is *methemoglobin,* and it is incapable of carrying oxygen.

One mole, or 32 grams, of oxygen combines with 16,700 grams of hemoglobin, and therefore the combining weight of hemoglobin is 16,700 grams. In all work on hemoglobin as a carrier of oxygen, 16,700 grams is considered to be the equivalent weight of hemoglobin. Therefore, 16,700 grams of hemoglobin is one equivalent, and 16.7 grams is one milliequivalent. A solution containing 16.7 grams per liter has a concentration of 1 milliequivalent per liter, or 1 mEq/l.

The molecular weight of hemoglobin is about 67,000, or four times the combining weight. For this reason we conclude that one molecule of hemoglobin actually combines with four molecules of oxygen according to the equation

$$4O_2 + Hb \quad \rightleftharpoons \quad Hb(O_2)_4. \tag{15}$$

However, this fact is ignored in acid-base chemistry, and equation (14) rather than equation (15) is used to express the combination of oxygen with hemoglobin.

One mole of oxygen at standard temperature and pressure occupies 22.4 cubic deciliters, or 22,400 cc. By the equation

$$\frac{22,400 \text{ cc per mole } O_2}{16,700 \text{ g per equivalent hemoglobin}}$$

$$= 1.34 \text{ cc } O_2 \text{ per g hemoglobin}, \tag{16}$$

we calculate that 1 g of hemoglobin combines with 1.34 cc of oxygen.

The average amount of hemoglobin in erythrocytes of 100 milliliters of blood is 15 grams. This amount of hemoglobin, when fully saturated, can carry 15(1.34) or 20.1 cc of oxygen. Fifteen grams of hemoglobin in 100 milliliters equals 150 grams in a liter. There is 150/16,700 or 0.0089 equivalent of hemoglobin per liter of normal blood. This quantity is customarily expressed in milliequivalents, and accordingly there are 8.9 milliequivalents of hemoglobin per liter of normal blood.

Example 4. A sample of blood is equilibrated at 38°C with a gas mixture having a P_{O_2} of 200 mm Hg, at which the hemoglobin is fully saturated. It is found to contain 21.3 vol % oxygen. How much hemoglobin is in the sample?

First Step. Oxygen dissolved in the blood must be subtracted from the total oxygen:

$$(O_2 \text{ dissolved}) = (0.023 \text{ cc per ml})(200 \text{ mm Hg})/760,$$
$$= 0.006 \text{ cc per ml},$$
$$= 0.6 \text{ vol } \%.$$

The total oxygen of the blood equals that dissolved plus that combined with hemoglobin:

$$(\text{Total } O_2) = (O_2 \text{ dissolved}) + HbO_2,$$
$$HbO_2 = 21.3 \text{ vol } \% - 0.6 \text{ vol } \%,$$
$$= 20.7 \text{ vol } \%.$$

Second Step.   Since 1.34 cc of oxygen combine with 1 g of hemoglobin,

g HbO_2 per 100 ml
$$= (20.7 \text{ cc per } 100 \text{ ml})/(1.34 \text{ cc per g}),$$
$$= 15.4 \text{ g } HbO_2 \text{ per } 100 \text{ ml}.$$

Since the equivalent weight of hemoglobin is 16,700,

Eq HbO_2 per liter $= (15.4 \text{ g per } 100 \text{ ml})(10)/(16,700),$
$$= 0.0093 \text{ Eq per liter},$$
$$= 9.3 \text{ mEq/l}.$$

The extent to which hemoglobin combines with oxygen depends upon the P_{O_2} of the solution. When the P_{O_2} is high, most or all of the hemoglobin is combined with oxygen; when the P_{O_2} is low, little hemoglobin is combined with oxygen. The extent of combination of hemoglobin with oxygen is discussed in terms of the *percentage saturation* of hemoglobin, which is the fraction of the total hemoglobin in the form of HbO_2 multiplied by 100:

$$\text{Percentage saturation} = 100 \ (HbO_2)/(HbO_2 + Hb). \qquad (17)$$

The relation between the percentage saturation and the P_{O_2} is expressed in the dissociation curve of hemoglobin, which is experimentally determined.

The dissociation curve of hemoglobin in whole blood with plasma pH of 7.40 at 38°C and at a P_{CO_2} of 40 mm Hg is plotted as the middle curve in figure 1. The P_{50}, or the P_{O_2} at which hemoglobin is 50% saturated, is 26 mm Hg.

If we know the percentage saturation of a sample of blood, and if all other variables affecting the shape of the dissociation curve are held constant, we can read the P_{O_2} from the dissociation curve, or if we know the P_{O_2} we can read the percentage saturation. The accuracy of the result depends upon the precision with which we measure the percentage saturation or the P_{O_2} and upon the correctness of the dissociation curve. The accuracy is especially poor when the upper end of the curve is used.

Example 5.   One hundred milliliters of a sample of blood containing 9.3 milliequivalents of hemoglobin per liter is equilibrated at 38°C with a gas mixture having a P_{O_2} of

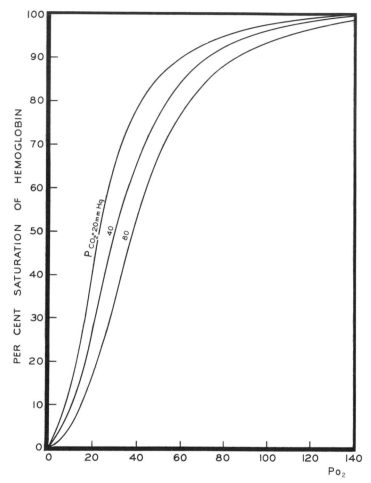

Fig. 1. The dissociation curve of hemoglobin at 38°C and at partial pressures of carbon dioxide equal to 20, 40, and 80 mm Hg.

200 mm Hg and a P_{CO_2} of 40 mm Hg. Another 100 ml of the same blood is equilibrated at the same temperature with a gas mixture having a P_{O_2} of zero and a P_{CO_2} of 40 mm Hg. The two samples are mixed. What is the P_{O_2} of the mixture?

First Step. The sample of oxygenated blood is the same as that described in example 4. It contains a total of 21.3 vol % oxygen of which 20.7 vol % is combined with hemoglobin and 0.6 vol % is dissolved. The other sample contains no oxygen. Consequently, the combined 200 ml contains 2(15.4) grams of hemoglobin and 21.3 cc of oxygen. Assuming, as a first approximation, that all the oxygen is combined with hemoglobin, the amount of HbO_2 is

$$\frac{21.3 \text{ cc O}_2}{1.34 \text{ cc O}_2 \text{ per gram HbO}_2} = 15.9 \text{ grams HbO}_2.$$

Percentage saturation
$$= 100 \ (15.9 \text{ g HbO}_2)/2(15.4 \text{ g HbO}_2 + \text{Hb})$$
$$= 52\% \text{ saturation}.$$

Second Step.   Reading from the middle curve in figure 1, the P_{O_2} is found to be approximately 30 mm Hg.

Third Step.   At this P_{O_2} the amount of oxygen dissolved is

(O$_2$ dissolved)
$$= 200(0.023 \text{ cc per ml})(30 \text{ mm Hg})/760$$
$$= 0.18 \text{ cc per 200 ml of blood}.$$

This is $100(0.18)/(21.3)$, or 0.8 per cent of the total oxygen; and the assumption that all oxygen is combined with hemoglobin in the mixed blood is better than 99% correct.

The shape of the dissociation curve depends mainly upon the P_{CO_2}, the pH, the concentration of organic phosphates, and the temperature.

An increase in the P_{CO_2} shifts the dissociation curve downward and to the right, and a decrease in the P_{CO_2} shifts the curve upward and to the left. Curves for P_{CO_2}'s of 80 and 20 mm Hg are also plotted in figure 1.

The process by which an increase in P_{CO_2} lowers the amount of oxygen combined with hemoglobin at a given P_{O_2} is called the Bohr effect.* It occurs because carbon dioxide combines with the α-amino groups of the N-terminal valines on the four polypeptide chains of hemoglobin. An increase in P_{CO_2} causes more carbon dioxide to react with these α-amino groups. This produces an allosteric effect in the hemoglobin molecule, and the reaction between the iron atoms and oxygen is shifted toward dissociation.

An increase in P_{CO_2} is accompanied by an increase in hydrogen ion concentration. This increase in $[H^+]$ also contributes to the Bohr effect. An increase in $[H^+]$ occurring without any concomitant change in P_{CO_2} also causes the dissociation curve to shift downward and to the right.

The shift in the dissociation curve of hemoglobin when the P_{CO_2} rises increases the amount of oxygen given up in the tissues. Arterial blood containing oxygen and having a P_{CO_2} of 40 mm Hg comes to the tissues where the P_{O_2} is low and the P_{CO_2} is about 46 mm Hg. The low P_{O_2} causes the reaction described in equa-

*It was described by Bohr, Hasselbalch, and Krogh, 1904, *Scand. Arch. Physiol.* 16:402. Laboratory gossip says most of the credit for the discovery belonged to Krogh.

tion (14) to shift to the left, and oxygen is released from oxyhemoglobin. Simultaneously, increase in P_{CO_2} causes the dissociation curve to shift downward and to the right. This means that at the P_{O_2} prevailing in the tissues hemoglobin combines with less oxygen than it would have combined with had the P_{CO_2} not increased. Therefore, oxyhemoglobin gives up an additional amount of oxygen to the tissues. Conversely, the fall in P_{CO_2} occurring as the blood passes through the lungs causes the dissociation curve to shift upward and to the left, and hemoglobin takes up an additional quantity of oxygen.

Organic phosphates combine with basic groups of hemoglobin. In the erythrocyte the chief organic phosphate is 2,3-diphosphoglycerate (2,3-DPG), which is produced by a reaction lying outside the normal glycolytic pathway. 2,3-DPG occurs in human erythrocytes at a concentration of about 0.8 moles per mole of hemoglobin. 2,3-DPG binds with the α-amino groups of the N-terminal valines of the β-chains (but not with those of the α-chains) and with the ϵ-amino groups of lysine. Upon binding with the α-amino groups of the β-chains 2,3-DPG has an effect similar to that of carbon dioxide when it reacts with the same groups: it causes hemoglobin to bind less oxygen. Therefore, an increase in the concentration of 2,3-DPG within the erythrocytes shifts the dissociation curve of hemoglobin downward and to the right. The concentration of 2,3-DPG rises when a person is acutely subjected to hypoxia. This makes more oxygen available to the tissues, for hemoglobin unloads more oxygen at a given P_{O_2}.

Adenosinetriphosphate also binds to hemoglobin, but in this respect it is less important than 2,3-DPG. Its concentration in the erythrocytes is only about a fifth that of 2,3-DPG.

Because carbon dioxide and 2,3-DPG combine with some of the same basic groups of hemoglobin, they compete with each other. A rise in the concentration of 2,3-DPG allows less carbon dioxide to combine with hemoglobin, and a rise in the P_{CO_2} to some extent displaces 2,3-DPG. On account of this competition their effects upon the hemoglobin dissociation curve are not simply additive. The shift in the curve produced by the two together is less than the sum of the effects of each separately.

The effects of carbon dioxide and 2,3-DPG are shown in figure 2. The dissociation curve of hemoglobin alone in a solution of pH 7.2, the pH of the interior of the erythrocytes, is shown at the left. The middle curve is that of the same hemoglobin solution at the same pH but with carbon dioxide at a P_{CO_2} of 40 mm Hg. The curve on the right is that obtained when 2,3-DPG was added at a ratio of 1.2 moles per mole of hemoglobin. The

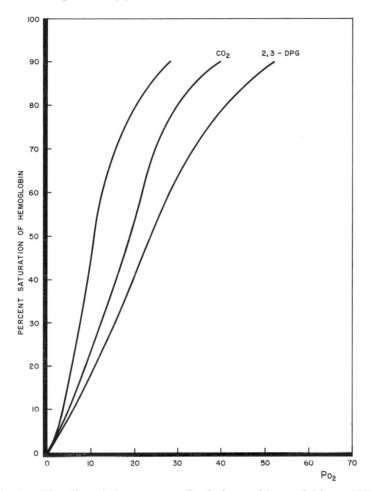

Fig. 2. The dissociation curves of solutions of hemoglobin at 37°C and pH 7.2. The left curve is that of hemoglobin alone. The center curve is that of hemoglobin after the addition of carbon dioxide at a partial pressure of 40 mm Hg. The right curve is that of hemoglobin to which was added 2,3-diphosphoglycerate. Curves constructed from unpublished data of L. Rossi-Bernardi. (Reproduced by permission.)

curve obtained when both carbon dioxide and 2,3-DPG are present is not shown, but it would be coincident with that of hemoglobin in whole blood shown in figure 1.

An increase in temperature shifts the hemoglobin dissociation curve downward and to the right, whereas a decrease in temperature shifts it upward and to the left. This is shown in figure 3. The temperature of exercising muscle rises several degrees, and the effect of the rise in temperature is to make hemoglobin unload more oxygen at a given P_{O_2}. On the other

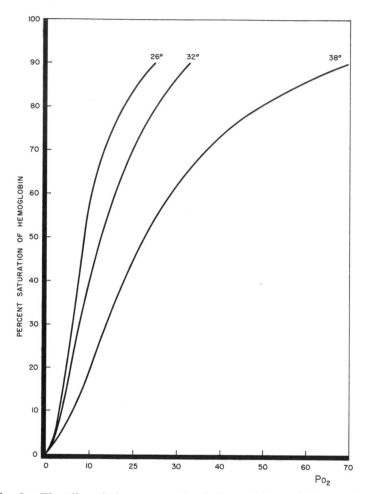

Fig. 3. The dissociation curve of solutions of hemoglobin at three different temperatures. Adapted from the data of Barcroft and King, 1909, *J. Physiol.* 39:374. (Reproduced by permission.)

hand, cold skin may be bright pink, not only because cold reduces the rate of oxygen consumption, but also because hemoglobin binds oxygen more tightly in the cold.

1.6. The pH Scale

The activity of hydrogen ions in a solution determines its acidity.

$$\text{Activity of hydrogen ions} = a_{H+}. \qquad (18)$$

The a_{H+} in pure water is about 10^{-7}. By convention, solutions having a_{H+} greater than 10^{-7} are acid solutions, and those having a_{H+} less than 10^{-7} are alkaline solutions.

In the human body a_{H+} ranges from about 0.13 in the most acidic gastric juice to about 0.000,000,03 in the most alkaline pancreatic juice. Expressed as powers of 10 these two activities are $10^{-.985}$ and $10^{-7.523}$ respectively. The pH scale was invented* to avoid the inconvenience of expressing such a wide range of activities or concentrations as powers of 10. The term pH was originally defined as the numerical value of the exponent, without regard for sign, of the concentration of hydrogen ions to the base 10. Thus, pH is the *power* (or *Potenz* in the original German) of $[H^+]_{10}$. Now the term is defined as the negative logarithm to the base 10 of the activity of hydrogen ions in solution. Its definition is expressed by the equation

$$pH = -\log a_{H+}. \tag{19}$$

In practice, activity of hydrogen ions in a solution is measured electrometrically. If the unknown solution is separated from a solution having a known, standard hydrogen ion activity by a membrane which is permeable only to hydrogen ions, the electrical potential E across the membrane is given by the equation

$$E = \frac{RT}{nF} \ln \frac{a_{H+} \text{ Standard}}{a_{H+} \text{ Unknown}}. \tag{20}$$

R is the gas constant; T is the absolute temperature; n is the valence (here equal to one); and F is the faraday. The logarithmic term is to the base e of natural logarithms and is 2.3 times logarithms to the base 10. The equation can be rewritten

$$E = 2.3 \frac{RT}{nF} \log (a_{H+} \text{ Standard})$$
$$- 2.3 \frac{RT}{nF} \log (a_{H+} \text{ Unknown}). \tag{21}$$

Since all values in the equation except E and $(a_{H+}$ Unknown$)$ are constant, the equation reduces to

$$E = A - b \log (a_{H+} \text{ Unknown}) \tag{22}$$

where A and b contain the constant values in equation (21). Equation (22) says that the potential across the membrane is a linear function of the logarithm of the unknown hydrogen ion activity. By definition

$$(pH \text{ Unknown}) = -\log (a_{H+} \text{ Unknown}). \tag{23}$$

Substituting this, we have

$$E = A + b \text{ (pH Unknown)}, \tag{24}$$

or

$$(pH \text{ Unknown}) = E/b - A/b. \tag{25}$$

The unknown pH is a linear function of the potential across the membrane.

*By Sorenson, Biochem. Z., 1909, 21:131, p. 134, where the term is written p_H.

Under appropriate conditions some kinds of glass behave as though they were membranes permeable only to hydrogen ions. This glass is formed into a bulb and filled with a standard solution having a known a_{H+}, and it is dipped into the solution whose unknown a_{H+} is to be measured. The circuit is completed by connecting a suitably prepared wire dipping into the fluid inside the bulb, or glass electrode, to one terminal of a potentiometer and by connecting a standard electrode, usually a calomel electrode, to the other terminal of the potentiometer. The calomel electrode is electrically connected with the unknown solution by a salt bridge, usually a saturated solution of potassium chloride. Then, if the apparatus is correctly constructed, the potential read is directly proportional to the pH of the unknown solution. The commercially available apparatus is usually standardized and calibrated so that its meter reads in terms of the pH of the unknown solution.

In an infinitely dilute solution the activity of hydrogen ions (a_{H+}) is exactly equal to the concentration of hydrogen ions $[H^+]$. In more concentrated solutions the activity is usually less than the concentration, and this fact is expressed by the equation

$$a_{H+} = f[H^+]. \tag{26}$$

Here f is the activity coefficient whose numerical value is less than one and which must be experimentally determined.

The activity coefficient for hydrogen ions in blood is not known; but because their concentration is so small, blood probably behaves as though it were an infinitely dilute solution. Therefore, the activity coefficient is probably close to one; and moreover, there is no reason to think that it varies under most physiological circumstances. Consequently, we will assume it to be one, and we write the equations

$$pH = -\log (a_{H+}), \tag{27}$$
$$= -\log (f[H^+]), \tag{28}$$
$$= -\log (1)[H^+], \tag{29}$$
$$= -\log [H^+]. \tag{30}$$

Gastric juice is the only body fluid for which the activity coefficient cannot be assumed to be nearly equal to one. In a sample of gastric juice whose pH is 1.00 and in which the sum of $[Na^+]$ and $[K^+]$ is 50 mEq/l, the activity coefficient of hydrogen ions is 0.810. Thus the activity of hydrogen ions is 0.100, but the concentration of hydrogen ions is 0.100/0.810, or 0.124 Eq/l.

Example 6.   The pH of normal arterial blood is 7.41. What is its hydrogen ion concentration?

The negative logarithm of the hydrogen ion activity is 7.41. Therefore, the logarithm of it is -7.41. This can also be written $0.59 - 8$.

The antilogarithm of 0.59 is 3.9, and the antilogarithm of -8 is 10^{-8}. The product of 3.9 and 10^{-8} is 3.9×10^{-8}.

Assuming the activity coefficient to be one, the hydrogen ion concentration is 3.9×10^{-8} moles per liter or 0.000,000,039 moles per liter.

For convenience the concentration of hydrogen ions in blood is often expressed in terms of nanomoles per liter. A nanomole is 10^{-9} moles. In this unit the result of example 6 is 39 nanomoles per liter, or 39 nM/l.

Example 7.   The extreme ranges of pH of arterial blood are 6.90 and 7.80. Calculate and tabulate the hydrogen ion concentrations corresponding to these limits and for the intermediate pH values in steps of 0.10 pH units.

For pH 6.90 $[H^+]$ = antilog (-6.90),
            = 1.26×10^{-7} moles per liter,
            = 126 nanomoles per liter.

For pH 7.80 $[H^+]$ = antilog (-7.80),
            $- 1.6 \times 10^{-8}$ moles per liter,
            = 16 nM/l.

Or   pH 6.90 = 126 nM/l
           7.00 = 100
           7.10 =   79
           7.20 =   63
           7.30 =   50
           7.40 =   40
           7.50 =   32
           7.60 =   25
           7.70 =   20
           7.80 =   16

The pH scale is a logarithmic one; and, as example 7 shows, equal intervals on the pH scale do not correspond to equal intervals on the hydrogen ion scale. The fact that pH is usually plotted on a linear scale distorts the physiological reality, which is the hydrogen ion concentration, lying behind the pH scale.

Example 8.   An increase of 25 nM/l in hydrogen ion concentration is frequently encountered, but a decrease of 25 nM/l is not compatible with life. What are the corresponding pH's?

Normal $[H^+]$ is 39 nM/l. An increase of 25 nM/l makes $[H^+]$ equal to 64 nM/l. The pH corresponding to this is 7.19.

A decrease of 25 nM/l makes $[H^+]$ equal to 14 nM/l. The pH corresponding to this is 7.85.

It is not correct to take arithmetical means of pH values. If a number is wanted to express the average pH, individual pH values must be converted to corresponding hydrogen ion concentrations. The arithmetical mean of these can be taken, and the mean can be reconverted to pH. It is also incorrect to speak of percentage changes in pH, for percentage changes are appropriate only to a linear scale, not a logarithmic one.

1.7. Buffer Action

An acid is a compound which is capable of giving off a hydrogen ion (or proton), and a base is one which is capable of accepting a hydrogen ion. A common acidic group is the carboxyl group (R—COOH). The structure of this group can be written

$$R-\overset{\overset{\displaystyle O}{\|}}{C}-OH.$$

The letter R represents any other group which may be attached to the carbon atom. When the compound is dissolved in water, the —OH group dissociates to give a hydrogen ion and an anion:

$$R-\overset{\overset{\displaystyle O}{\|}}{C}-OH \rightleftharpoons R-\overset{\overset{\displaystyle O}{\|}}{C}-O^- + H^+. \tag{31}$$

The arrows indicate that the reaction is reversible. Since the anionic product of the reaction is capable of accepting a hydrogen ion as the reaction procedes to the left, the anion is a base. Such an anion is called the conjugate base of the corresponding acid.

Another group capable of giving off a hydrogen ion is the ammonium group ($-NH_3^+$). The reaction is

$$R-NH_3^+ \rightleftharpoons R-NH_2 + H^+. \tag{32}$$

In this instance, the compound $R-NH_2$ is the conjugate base, and the cationic $R-NH_3^+$ is the acid.

If hydrogen ions are added to a solution containing an acid and its conjugate base, the increase in hydrogen ion concentration drives the reaction depicted in these equations to the left. Some hydrogen ions combine with the conjugate base to form the acid, and therefore some hydrogen ions disappear from the solution. The final concentration of hydrogen ions after the addition of acid is lower than it would have been if the conjugate base had not been present. On the other hand, a decrease in the hydrogen ion concentration of the solution drives the reactions to the right, and some of the acid molecules give up hydrogen ions. The final concentration of hydrogen ions in the solution is higher than it would have been if the acid had not been present. Because an acid and its conjugate base tend to minimize changes in hydrogen ion concentration of a solution, the pair act as a *buffer*.

The action of a buffer can be described quantitatively in terms of its *titration curve*. A titration curve is constructed by dissolving a known amount of buffer in water and measuring the pH of the solution. Then a known amount of acid is added or removed, and the pH is again measured. The process is repeated until the entire buffering range of the buffer is covered. When the amounts of acid added are plotted as ordinates against the pH's as abscissas, the titration curve is constructed. The curve of a representative carboxyl buffer is shown in figure 4.

The buffer solution represented in figure 4 was made up so that it contained 10 millimoles of the buffer. It was adjusted so that half the buffer, or 5 millimoles, was in the undissociated acidic form (R—COOH) and the other half was in the disso- ciated basic form (R—COO⁻). The pH of the solution was found to be 4.73. This is represented by point A. When 2 millimoles of acid were added to the solution the pH was found to be 4.36 as represented by point B. The pH had changed −0.37 pH units. In a similar fashion other points on the buffer curve were found by adding or removing hydrogen ions, and the whole curve was drawn.

The *buffer value* of a solution is the quantity of hydrogen ions which can be added to, or removed from, a solution with a change in one pH unit. This value is given by the slope of the titration curve. The titration curve shown in figure 4 is not a

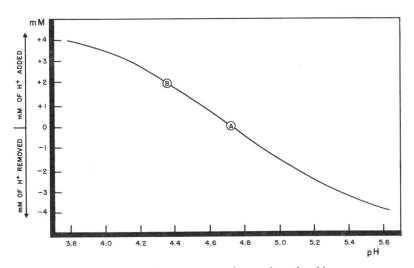

Fig. 4.   Titration curve of a representative carboxyl acid.

straight line, and the slope changes as the pH changes. The true slope of the curve at any point is the slope of a straight line tangent to the curve at that point. However, as a first approximation, the curve between points A and B can be considered to be straight. Between these two points addition of 2.0 millimoles of acid caused a fall of 0.37 pH units. Therefore, the buffer value of the solution between points A and B is approximately 2.0 millimoles of acid per -0.37 pH units. By the proportionality equation

$$\frac{2.0 \text{ mM H}^+}{-0.37 \text{ pH units}} = \frac{5.4 \text{ mM H}^+}{-1 \text{ pH unit}} \tag{33}$$

it can be seen that the buffer value of the solution is about -5.4 millimoles of acid per pH unit between points A and B.

The buffer value of this buffer is different at other points on the curve, for the slope of the curve changes. The slope is greatest at the middle of the curve where half the buffering groups are in the undissociated acidic state and half are in the dissociated basic state. The pH at the point where the concentration of two members of a conjugate pair is equal is called the pK of the buffer.

1.8. Hemoglobin as a Buffer: The Titration Curve of Oxyhemoglobin

A protein such as hemoglobin is a buffer because its molecule contains a large number of acidic or basic groups: carboxyl ($-COOH$), amino ($-NH_2$), ammonium ($-NH_3^+$), or guanidino ($-NH-CNH-NH_2$). There may be other types of buffering groups such as the imidazole group of histidine. At the isoionic point of a protein the number of anionic groups equals the number of cationic groups, and the net charge on the protein is zero.

A protein can be represented diagrammatically as in figure 5. The isoionic protein is shown as having four acidic and four basic groups. When four hydrogen ions are added to the solution, three of the hydrogen ions combine with the basic carboxyl groups, suppressing their ionization and forming undissociated groups. One of the hydrogen ions remains in solution, making it more acidic. Had the protein not been present, all four hydrogen ions would have remained in solution, and the solution's acidity would have been greater. Conversely, when hydroxyl ions are added to the solution, some of the acidic groups give up their hydrogen ions which combine with the hydroxyl groups to form water. The decrease in acidity of the solution is not so great as it would have been had the protein not been present. In this way protein buffers the solution, preventing or reducing shifts in

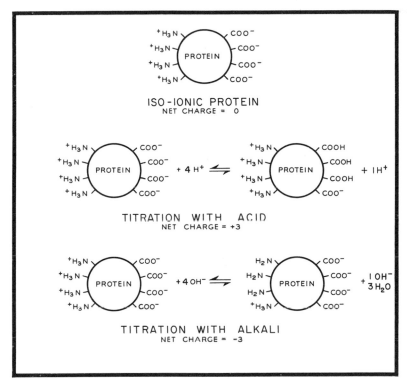

Fig. 5.   Schematic representation of the buffering action of a protein.

hydrogen ion concentration when acid is added to or removed from solution.

Much of the buffering by oxyhemoglobin in the physiological range is done by imidazole groups, particularly the C-terminal ones of the β-chains. These groups undergo the changes shown in figure 6. The α-amino groups of the α-chains which are not combined with carbon dioxide or with 2,3-DPG also buffer in the physiological range. Some potential buffer groups are not available, because they are buried within the hemoglobin molecule or because their pK's are too low or too high.

The buffering power of a protein is expressed in its titration curve. The titration curve of oxyhemoglobin is shown in figure 7. In order to obtain this particular curve a solution of human hemoglobin was equilibrated at 37°C with a gas mixture containing oxygen at a P_{O_2} of 600 mm Hg and carbon dioxide at a P_{CO_2} of 39 mm Hg. The pH of the solution was measured. Then a known amount of a standard solution of potassium carbonate was added to remove acid from the solution. The sample was again equilibrated at 37°C with the same gas mixture, and after equilibration its pH was again measured. In order to keep the curve

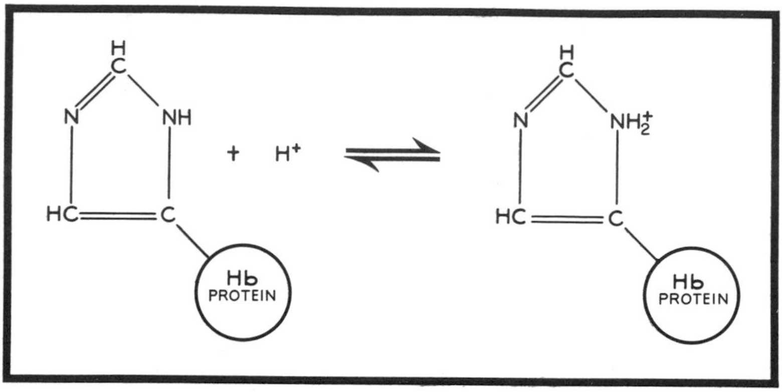

Fig. 6.   Schematic representation of the buffering action of the imid-azole group of hemoglobin.

consistent with the rest of the buffering curves used in this book, the data have been presented in terms of acid *added* to the solution instead of in terms of acid removed, as was the case in the actual experiment.

Over the physiological range of pH the titration curve of oxyhemoglobin is almost a straight line. The slope of the central part of the curve drawn in figure 7 is -7.2 millimoles of acid per milliequivalent of hemoglobin per pH unit.

Example 9.   One liter of human oxyhemoglobin solution held at 37°C and equilibrated with a gas mixture containing carbon dioxide at a P_{CO_2} of 39 mm Hg contains 8.7 milliequivalent of hemoglobin. The pH is 7.24. Ten millimoles of hydrochloric acid are added. What is the pH of the final solution?

If the addition of 7.2 millimoles of acid to a similar solution containing one milliequivalent of hemoglobin per liter causes the pH to fall one unit, the addition of 10 millimoles of acid to 8.7 milliequivalents will cause the pH to fall according to the equation

$$\frac{7.2 \text{ mM acid}}{(-1 \text{ pH unit})(1 \text{ mEq HbO}_2)}$$
$$= \frac{10 \text{ mM acid}}{(x \text{ pH units})(8.7 \text{ mEq HbO}_2)},$$
$$= -10/(8.7)(7.2),$$
$$= -0.16 \text{ pH units.}$$

The final pH is $(7.24 - 0.16)$ or 7.08.

What would be the pH change if oxyhemoglobin were not present?

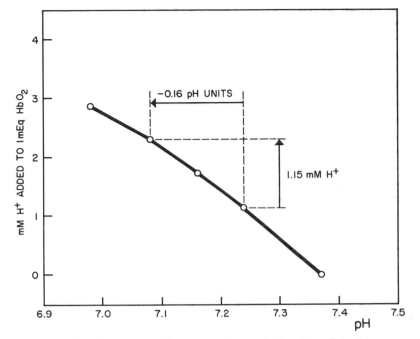

Fig. 7.   Titration curve of human oxyhemoglobin at 37°C in the presence of carbon dioxide at a partial pressure of 39 mm Hg. Adapted from the data of Rossi and Roughton, 1967, *J. Physiol.* 189:1. (Reproduced by permission.)

The addition of 10 millimoles of acid to 1 liter neutral unbuffered solution would bring the hydrogen ion concentration to 10 millimoles per liter, or 0.010 M. If the activity coefficient of hydrogen ions is one at this dilution, the pH equals the negative logarithm of the hydrogen ion concentration. This is pH 2.

1.9. The Direct Combination of Carbon Dioxide with Hemoglobin: Carbamino Compounds

Between 5 and 15 per cent of carbon dioxide carried from tissues to lungs in a resting man is carried as carbamino compounds.

Carbon dioxide reacts with amino groups to form carbamino compounds according to the equation

$$R—NH_2 + CO_2 \;\rightleftharpoons\; R—NHCOO^- + H^+. \tag{34}$$

The reaction occurs very rapidly and does not require a catalyst. Carbon dioxide does not react with $—NH_3^+$ groups to form carbamino compounds.

Carbon dioxide reacts with the N-terminal α-amino groups on all four chains of hemoglobin. It does not combine with any other groups on the hemoglobin molecule. 2,3-DPG also reacts with the same α-amino groups on the β-chains but not with those of the α-chains. Therefore, there are two different types of amino groups with which carbon dioxide reacts: those on the α-chains where it does not compete with 2,3-DPG and those on the β-chains where it does compete.

The amount of carbon dioxide carried by hemoglobin in the oxygenated and deoxygenated forms is shown in figure 8. As the pH increases and the solution becomes more alkaline, the fraction of α-amino groups in the $-NH_2$ form increases. Therefore, there are more groups with which carbon dioxide can combine, and the amount of carbon dioxide carried as carbamino compounds increases with increasing pH. More carbon dioxide is carried by deoxygenated hemoglobin than by oxygenated hemoglobin. At pH 7.2, the usual pH of the interior of the erythrocytes, the deoxygenation of one equivalent of hemoglobin results in an increase of 0.081 equivalents of carbamino compounds.

Example 10. If $\Delta HbCO_2/\Delta HbO_2$ is 0.081, and if the respiratory quotient is 0.95, what fraction of the carbon dioxide produced in the tissues is carried to the lungs as carbamino compounds?

First Step. A respiratory quotient of 0.95 means that 0.95 volumes of carbon dioxide are produced for every

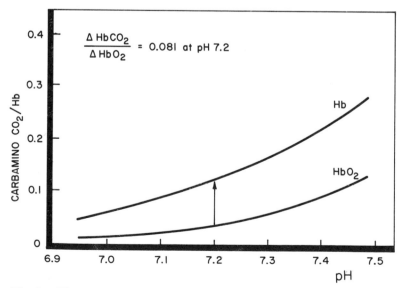

Fig. 8. The amount of carbon dioxide carried as carbamino compounds within erythrocytes when the hemoglobin is either completely oxygenated or completely deoxygenated. Adapted from Bauer and Schroeder, 1972, *J. Physiol.* 227:457. (Reproduced by permission.)

volume of oxygen used. Since equal volumes of gases contain almost equal numbers of molecules, 0.95 moles of carbon dioxide are produced per mole of oxygen used when the respiratory quotient is 0.95.

Second Step.   If we neglect the very small change in dissolved oxygen occurring during the respiratory cycle, we can say that for every mole of oxygen used, one equivalent of hemoglobin is deoxygenated. The deoxygenation of one equivalent of hemoglobin results in an increase of 0.081 moles of carbamino compounds.

Third Step.   For every mole of oxygen used, 0.95 moles of carbon dioxide are produced and 0.081 moles of this forms carbamino compounds. The fraction of the carbon dioxide produced carried as carbamino compounds is 0.081/0.95 or 0.085.

In this example, carbamino compounds account for the carriage of 8.5% of the carbon dioxide produced.

The fact that a change in the state of oxygenation of hemoglobin affects the amount of carbon dioxide carried as carbamino $-CO_2$ implies that the amount of carbon dioxide carried as carbamino$-CO_2$ affects the oxygen carrying capacity of hemoglobin. Section 1.3 shows that the hemoglobin dissociation curve is in fact displaced downward and to the right by an increase in P_{CO_2}. This effect of carbon dioxide on hemoglobin is largely if not entirely mediated by carbamino formation.

Equation (34) shows that when carbon dioxide combines with a $-NH_2$ group, the resulting compound ionizes to give one hydrogen ion and one ionized carboxyl group. The carboxyl group is a relatively strong acid, and its pK is below 5.8. Consequently, the carboxyl group plays no part in buffering in the physiological range.

Equation (34) implies that for every millimole of carbamino$-CO_2$ formed, one millimole of hydrogen ions is released. It is probable that somewhat more than one millimole of hydrogen ions is released when a millimole of carbamino$-CO_2$ is formed. The reason is that the amino groups of hemoglobin also participate in the equilibrium expressed in the equation

$$H^+ + R-NH_2 \rightleftharpoons R-NH_3^+. \tag{35}$$

When carbamino compounds are formed by the reaction between CO_2 and $R-NH_2$ the concentration of $R-NH_2$ is reduced, and the reaction described in equation (35) proceeds to the left. For every millimole of $R-NH_2$ produced by this reaction, an additional millimole of hydrogen ions is released. The position of the equilibrium described in equation (35) is such that when one millimole of carbamino—CO_2 is formed in blood at a pH of 7.4, 1.5 millimoles of H^+ are given off by hemoglobin. Of this, one millimole is produced by the reaction in equation (34), and the rest is produced by the reaction in equation (35).

1.10. Hemoglobin as a Buffer: The Effect of Reduction

The total quantity of carbon dioxide carried in blood depends, in part, upon the state of oxygenation of hemoglobin. If blood containing deoxygenated hemoglobin is oxygenated at a constant P_{CO_2}, the oxygenated blood contains less carbon dioxide than it did in the deoxygenated state. Oxygenation drives out carbon dioxide. Conversely, deoxygenation of blood allows it to take up more carbon dioxide at a constant P_{CO_2}. The change in carbon dioxide content of blood accompanying oxygenation or deoxygenation of hemoglobin is called the Haldane effect.*

The Haldane effect depends upon the change in the nature of hemoglobin as a buffer when it is oxygenated or reduced. When hemoglobin is oxygenated, it becomes a stronger acid and gives off hydrogen ions. These hydrogen ions combine with bicarbonate ions to form carbonic acid which in turn is dehydrated to carbon dioxide and water. At a constant P_{CO_2} the newly formed carbon dioxide passes into the gas phase and the total carbon dioxide content of the blood falls. When oxygenated hemoglobin is deoxygenated, it becomes a weaker acid and removes hydrogen ions from solution. The fall in concentration of hydrogen ions causes some carbonic acid to ionize, and the fall in carbonic acid concentration allows more carbon dioxide to enter the blood from the gas phase. Thus the total carbon dioxide content of blood rises as hemoglobin is deoxygenated.

This change in the acidic properties of hemoglobin occurs because the hemoglobin molecule contains ionizing groups whose strength changes with the state of oxygenation. One of these groups belongs to the C-terminal histidines in the β-chains of hemoglobin. In oxygenated hemoglobin these particular histidine groups are free in solution. There they have the normal pK to be expected of imidazole groups. When oxygenated hemoglobin is deoxygenated, the β-chains change their shape, and in their new conformation the C-terminal histidines react with the aspartates at position 94 in the same chain. This interaction raises the apparent pK of the imidazole group of the histidines,

*The effect was first described by Christiansen, Douglas, and Haldane, 1914, *J. Physiol.* 48:244. The Haldane effect is complimentary to the Bohr effect, and the existence of the Haldane effect was implied by the discovery of the Bohr effect. Discovery of the Haldane effect had to wait development of improved methods of gas analysis.

and hydrogen ions are taken up from solution. When deoxygenated hemoglobin is again oxygenated, the C-terminal histidines become free in solution once more; their pK falls, and they give off hydrogen ions. This change in the properties of the imidazole groups is shown schematically in figure 9.

Instead of saying that the imidazole group becomes a stronger acid when hemoglobin is oxygenated, one could say that it becomes a weaker base. The terminology is unimportant so long as the process is understood.

The α-amino groups of the N-terminal valines of the α-chains also participate in the Haldane effect. Oxygenation of hemoglobin lowers their pK; they become stronger acids and give off hydrogen ions. Deoxygenation raises their pK, and they take up hydrogen ions.

The effect of reducing hemoglobin is shown in figure 10, which contains the titration curve of oxyhemoglobin given in figure 7. In order to obtain the titration curve of reduced hemoglobin, a sample of the same blood was equilibrated at 37°C with a gas mixture containing no oxygen but containing carbon dioxide at a P_{CO_2} of 39 mm Hg. Because the gas mixture contained no oxygen, oxyhemoglobin was reduced. The pH of the solution was measured. Then a known amount of standard potassium carbonate was added to the reduced blood to remove acid from the solution. A sample was again equilibrated with the same gas mixture at 37°C, and after equilibration its pH was again measured. In order to keep the curve consistent with the rest of the blood buffering curves used in this book, the data have been

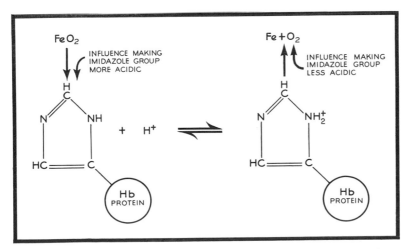

Fig. 9.   Schematic representation of the effect of oxygenation and reduction upon the buffering action of the imidazole group of hemoglobin.

presented in terms of acid added to the solution instead of in terms of acid removed as was the case in the actual experiments. The points for reduced hemoglobin are plotted as open circles in figure 10.

The horizontal arrow in figure 10 shows the shift in pH occurring when hemoglobin is reduced without any change in the amount of acid added or removed by external means. This rise in pH is caused by removal of hydrogen ions from solution by hemoglobin as it becomes a weaker acid upon removal of oxygen from its iron atoms. On the average the shift in pH when hemoglobin is changed under these conditions from fully oxygenated to fully reduced is 0.048 pH units.

At the same time hydrogen ions are removed from the solution by hemoglobin, an equal number could be added to the solution from the outside. Addition of extra hydrogen ions would prevent any rise in pH of the solution resulting from reduction of oxyhemoglobin. This is shown by the vertical arrow *1* which represents the change occurring when oxyhemoglobin is reduced

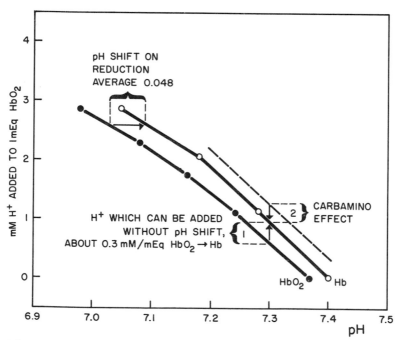

Fig. 10.   Titration curves of oxyhemoglobin (filled circles) and of reduced hemoglobin (open circles) at 37°C in the presence of carbon dioxide at a partial pressure of 39 mm Hg. The dashed line parallel to the curves shows the approximate position of the titration curve of reduced hemoglobin in the absence of carbon dioxide. Adapted from the data of Rossi and Roughton, 1967, *J. Physiol.* 189:1. (Reproduced by permission.)

and acid is simultaneously added. Measurement of the vertical distance between the two curves shows that for every milliequivalent of oxyhemoglobin reduced, approximately 0.3 millimoles of acid can be added, and the pH will remain constant.

The measurements upon which figure 10 is based were made on hemoglobin solutions whose P_{CO_2} was kept constant at the physiological value of 39 mm Hg. Section 1.9 shows that when oxyhemoglobin is reduced it forms carbamino compounds and that when carbamino compounds are formed, hydrogen ions are given off. Therefore, two antagonistic processes occur when oxyhemoglobin is reduced in the presence of carbon dioxide: (1) reduction of oxyhemoglobin makes the protein a weaker acid, and hydrogen ions are taken up from the solution; but (2) reduction of oxyhemoglobin allows formation of more carbamino —CO$_2$, and hydrogen ions are given off. In the pH range of 6.9 to about 7.5, the first process takes up more hydrogen ions than are given off by the second. Therefore, either the pH of hemoglobin solutions rises when the hemoglobin is reduced, or acid can be added when hemoglobin is reduced with no change in pH.

If oxyhemoglobin is reduced in the *absence* of carbon dioxide, formation of carbamino compounds cannot occur, and no hydrogen ions are given off by that process. If no acid is added to the solution, the pH increase is very much greater than that shown in figure 10. If acid is simultaneously added to prevent the rise in pH, more acid can be added. The titration curve which would be obtained upon reduction of oxyhemoglobin in absence of carbon dioxide is approximately represented by the dashed line parallel to the other titration curves in figure 10. Arrow *2*, representing the vertical distance between the dashed line and the titration curve of reduced hemoglobin, shows the magnitude of the antagonistic carbamino effect.

1.11. Carriage of Carbon Dioxide in the Blood: Qualitative Aspects

Arterial blood coming to the tissues contains a large amount of oxyhemoglobin and a minimum amount of carbon dioxide. In passing through the tissues, oxyhemoglobin gives up oxygen to the tissues, and the tissues deliver carbon dioxide to the blood. Carbon dioxide produced by the tissues is probably released into the blood in the form of carbon dioxide dissolved in water. Carbon dioxide diffuses into the plasma where three things happen to it:

1. The largest fraction of carbon dioxide diffuses through the plasma into the erythrocytes where buffering mechanisms are available to deal with it.

2. Dissolved carbon dioxide forms some carbamino compounds with plasma proteins. Since there are relatively few amino groups on plasma proteins capable of combining with carbon dioxide, a total of not more than 0.5 millimole of carbon dioxide is carried in a liter of plasma as carbamino—CO_2. Carbamino compounds of plasma are not affected by the state of oxygenation of the blood, and the amount of carbamino—CO_2 of plasma does not change significantly as the blood becomes venous.

3. The rest of the dissolved carbon dioxide remains in the plasma. Dissolved carbon dioxide reacts with water according to the equation

$$CO_2 + H_2O \rightleftharpoons H_2CO_3. \tag{36}$$

Combination of carbon dioxide with water to form carbonic acid is called *hydration* of carbon dioxide, and the reverse reaction is called *dehydration*. The equilibrium of the reaction is far to the left, and in plasma the concentration of dissolved carbon dioxide is about 1,000 times greater than the concentration of carbonic acid. The increase in concentration of dissolved carbon dioxide occurring when blood becomes venous drives the reaction slightly to the right, and a minute amount of carbon dioxide is hydrated to form carbonic acid. The small amount of carbonic acid formed ionizes according to the equation

$$H_2CO_3 \rightleftharpoons H^+ + HCO_3^-. \tag{37}$$

The hydrogen ions produced by the ionization of carbonic acid are buffered by the weak buffering system of the plasma with an attendant slight fall in pH, and the bicarbonate ions remain in the plasma. These reactions are summarized in figure 11.

   Carbon dioxide diffusing into the erythrocytes is carried in three ways:

1. Some remains in the erythrocytes as dissolved carbon dioxide.
2. A significant fraction of the carbon dioxide combines with hemoglobin to form carbamino—CO_2. As oxyhemoglobin is reduced by giving up oxygen to the tissues it becomes capable of combining with an increased amount of carbon dioxide. When carbamino compounds are formed with hemoglobin, hydrogen ions are given off as described in section 1.9. The hydrogen ions are buffered within the erythrocytes by hemoglobin.
3. The largest fraction of carbon dioxide entering the erythrocytes is hydrated to form carbonic acid. In turn, most of the

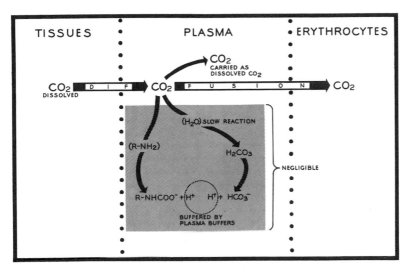

Fig. 11.   Schematic representation of the processes occurring when carbon dioxide passes from tissues to plasma.

newly formed carbonic acid ionizes to give hydrogen ions and bicarbonate ions. These reactions proceed because two processes remove the reaction products from the erythrocytes as soon as they are formed: hemoglobin buffers most of the hydrogen ions, and much of the bicarbonate diffuses into the plasma.

Some of the hydrogen ions arising from carbamino—CO_2 formation and from ionization of carbonic acid combine with the buffering groups of the hemoglobin molecule, titrating them in the acid direction. This process is shown as the arrow from point A to point B in figure 12. The titration of hemoglobin is accompanied by a small fall in the pH of the blood.

At the same time carbon dioxide is being delivered by the tissues to the blood, some oxyhemoglobin is being reduced and reduction of hemoglobin makes it a weaker acid. Consequently, hydrogen ions are taken up by reduced hemoglobin. This process is shown in figure 12 by the arrow from point B to point C which represents the amount of hydrogen ions taken up when one milliequivalent of oxyhemoglobin is 70 per cent reduced.

Both processes—titration of hemoglobin and its reduction—occur simultaneously, and the actual buffering of hydrogen ions is represented by the arrow from point A to point C.

The law of electrical neutrality of solutions applies within the erythrocytes. There must be the same number of positive

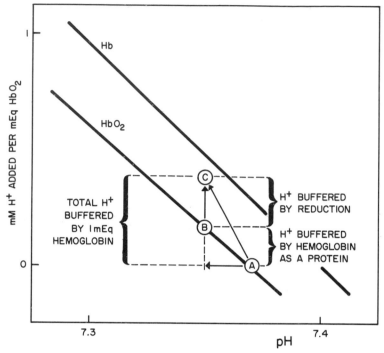

Fig. 12. Graphical representation of the buffering of hydrogen ions by hemoglobin when arterial blood passes through the tissues and oxyhemoglobin is partially reduced. Curves constructed from the data of Rossi and Roughton, 1967, *J. Physiol.* 189:1. (Reproduced by permission.)

ionic charges in solution as there are negative charges. Hemoglobin, before it buffers the hydrogen ions, has a certain number of net negative charges. These negative charges are balanced by positive charges of cations within the erythrocytes, the cations being chiefly potassium and sodium ions.

When carbonic acid ionizes it forms an equal number of hydrogen cations and bicarbonate anions. The hydrogen ions combine with hemoglobin, and the net negative charge on hemoglobin is reduced. Potassium and sodium ions are then balanced electrically against bicarbonate anions, and electrical neutrality of the solution is maintained.

Bicarbonate ions within the erythrocytes are in equilibrium with bicarbonate ions in the plasma. As the result of the changes which occur when arterial blood becomes venous blood, the bicarbonate concentration of the erythrocytes increases, and the bicarbonate concentration of the erythrocytes is no longer in equilibrium with that of plasma. Consequently, bicarbonate ions diffuse from erythrocytes into plasma. Since bicarbonate ions are negatively charged, electrical neutrality of erythrocytes and plasma would be disturbed unless one of two things happened: either an equal number of positively charged cations could also diffuse from erythrocytes into plasma, or an equal number of negatively charged anions could diffuse from plasma into erythrocytes. The membrane of the erythrocytes is impermeable to cations, at least over the very brief time in which these exchanges occur, and inward diffusion of anions takes place. The anions available in plasma are chloride ions, and chloride ions diffuse into erythrocytes as bicarbonate ions diffuse out. This exchange—the chloride shift—continues until equilibrium is reached. As its result, the concentration of bicarbonate in plasma increases, and a large part of carbon dioxide added to venous blood is carried in the plasma. This is the result, not of the weak buffering power of plasma, but of the strong buffering power of hemoglobin within the erythrocytes.

Plasma and erythrocytes are in osmotic equilibrium. This means that each volume of water in plasma contains the same number of osmotically active particles as does an equal volume of water in erythrocytes. The osmotically active particles are chiefly the small ions such as sodium, potassium, chloride, and bicarbonate ions. Plasma proteins and hemoglobin, being large molecules, have slight osmotic activity, and only 0.4 per cent of the osmotic pressure of plasma or erythrocytes is the result of the presence of these molecules. When carbon dioxide is added to blood and goes through the reactions described above, one result is that the net number of negative charges on hemoglobin is reduced, and the negative charges are replaced by chloride and bicarbonate ions. These ions are osmotically active, while the hemoglobin charges they replace have negligible osmotic activity. As a result, the total osmotic pressure of the interior of erythrocytes increases, and erythrocytes are no longer in osmotic equilibrium with plasma. In order that equilibrium may be restored, water moves from plasma into erythrocytes. Consequently, erythrocytes swell slightly as arterial blood becomes venous.

The changes occurring in erythrocytes as they buffer carbon dioxide are shown in figure 13, which, together with figure 11, describes qualitatively events taking place in tissue capillaries. The reverse processes occur in the lungs as blood gives off carbon dioxide and takes up oxygen.

The rates at which all these reactions proceed have no effect upon the equilibrium finally reached, for the equilibrium is the same whether it is reached rapidly or slowly. However, the rate of circulation of the blood sets time limits within which

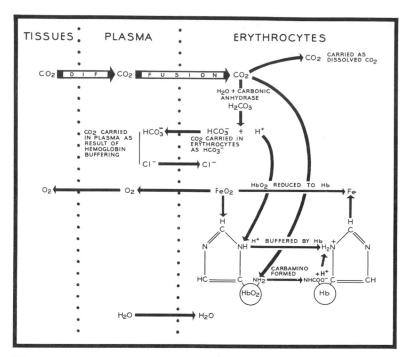

Fig. 13. Schematic representation of the processes occurring when carbon dioxide passes from tissues into erythrocytes.

the reactions must occur. Erythrocytes spend less than one second in the capillaries of the lungs, and during this brief period the reactions which liberate carbon dioxide from venous blood into alveolar air must take place. All the reactions, with one exception, are very rapid. The exception is the hydration and dehydration of carbon dioxide described in equation (36). In the absence of a catalyst, this reaction is a slow one. If the uncatalyzed rate of dehydration of carbonic acid were the limiting reaction in the series, the liberation of carbon dioxide from erythrocytes in the lungs would require 100 seconds to reach 90 per cent of completion.

Almost all the carbon dioxide which is hydrated or dehydrated when blood takes up or gives off carbon dioxide undergoes those processes within erythrocytes where hemoglobin is present. Consequently, when carbon dioxide is hydrated, hydrogen ions liberated by subsequent ionization of carbonic acid are readily buffered; and, when it is dehydrated, hydrogen ions which must combine with bicarbonate ions are available from hemoglobin. Erythrocytes, but not plasma, contain a high concentration of the enzyme carbonic anhydrase which catalyses hydration and dehydration of carbon dioxide. As a result of the presence of the enzyme, the reaction within erythrocytes occurs very rapidly, and carbon dioxide is taken up in the tissues or liberated in the lungs in the time allowed by the rate of circulation of the blood. Carbonic anhydrase, like all other enzymes, merely accelerates the rate at which equilibrium is reached. Equilibrium states and buffering reactions occurring in blood would be the same were carbonic anhydrase absent, but in that case reactions depending upon hydration or dehydration of carbon dioxide could not occur in the brief time erythrocytes spend in the capillaries of the tissues and lungs.

1.12. Carriage of Carbon Dioxide in the Blood: Quantitative Aspects

The data in table 1 show the distribution of carbon dioxide in samples of arterial and venous blood taken from a normal man (A.V.B.) at rest.

The first part of the table shows that arterial blood contains 21.53 millimoles of carbon dioxide per liter. Venous blood contains 23.21 millimoles per liter; and the difference, 1.68 millimoles, is the amount carried from tissues to lungs by one liter of blood.

Of the one liter of blood, 60 per cent is plasma and 40 per cent is erythrocytes. The 600 milliliters of plasma of arterial blood contain a total of 15.94 millimoles of carbon dioxide, and the 600 milliliters of plasma of venous blood contain a total of 16.99 millimoles. The difference, 1.05 millimoles, is the amount of carbon dioxide carried in plasma from tissues to lungs. This is 62 per cent of the total amount carried. Of this 1.05 millimoles, only 0.09 millimole is carried as dissolved carbon dioxide, and the remaining 0.96 millimole is carried as bicarbonate ions.

The 400 milliliters of erythrocytes in one liter of blood carry 0.63 millimole of carbon dioxide from tissues to lungs. Of this 0.63 millimole, only 0.05 millimole is carried as dissolved carbon dioxide, 0.40 millimole as bicarbonate ions and the rest as carbamino—CO_2.

These data show that the major fraction of carbon dioxide is carried in plasma. Nevertheless, hemoglobin within the erythrocytes is responsible for carriage of most of the carbon dioxide.

The amount of carbonic acid formed in plasma and buffered by plasma proteins can be estimated from the data on plasma pH and the change in the number of negative charges on the proteins. Plasma proteins are titrated in the acid direction as the pH falls by 0.026 pH unit, and they take up a total of 0.09 millimole of hydrogen ions. These hydrogen ions come from carbonic acid

Table 1

Distribution of Carbon Dioxide in One Liter of Normal Human Blood Containing 8.93 Milliequivalents of Hemoglobin per Liter and Having a Hematocrit of 40 Per Cent. The Respiratory Quotient is 0.82.

	Arterial	Venous	Difference
Whole blood, 1 liter			
P_{O_2}, mm Hg	96	40	-56
P_{CO_2}, mm Hg	40	46	$+6$
Total O_2, millimoles	8.65	6.60	-2.05
Total CO_2, millimoles	21.53	23.21	$+1.68$
Plasma, 600 milliliters			
Total CO_2, millimoles	15.94	16.99	$+1.05$
Dissolved CO_2, millimoles	0.71	0.80	$+0.09$
Bicarbonate, millimoles	15.23	16.19	$+0.96$
pH	7.455	7.429	-0.026
Net negative charges on plasma proteins, millimoles	7.89	7.80	-0.09
Chloride, millimoles	59.59	58.72	-0.87
Erythrocytes, 400 milliliters			
Total CO_2, millimoles	5.59	6.22	$+0.63$
Dissolved CO_2, millimoles	0.34	0.39	$+0.05$
Carbamino-CO_2, millimoles	0.64	0.82	$+0.18$
Bicarbonate, millimoles	4.61	5.01	$+0.40$
Net negative charges on hemoglobin, millimoles	22.60	21.44	-1.16
Chloride, millimoles	18.11	18.98	$+0.87$

SOURCE: Data on blood of A.V.B. adapted from L. J. Henderson, *Blood* (Yale University Press: New Haven, 1928). (Reproduced by permission.)

formed in the plasma, and the bicarbonate ions which result from formation and buffering of carbonic acid are carried in the plasma. However, a total of 0.96 millimole of bicarbonate is carried in the plasma. The difference between 0.09 and 0.96 millimole, or 0.87 millimole, is carried in the plasma because hydrogen ions formed at the same time are buffered within the erythrocytes.

The hydrogen ion concentration of the arterial plasma, calculated from the pH, is 35, and that of venous plasma is 37 nanomoles per liter. Therefore, in changing from arterial to venous blood, the quantity of hydrogen ions in the 600 milliliters of plasma increased less than 2 nanomoles. At the same time the total carbon dioxide content of the plasma increased 1,050,000 nanomoles. Comparison of the two figures demonstrates the effectiveness of the buffers of the blood.

1.13. Fundamental Equations

In the following quantitative description of the carriage of carbon dioxide in blood a conventional system of symbols will be used. All quantities enclosed in square brackets represent concentrations. The subscripts p and c refer to concentrations in plasma and erythrocytes, respectively. Thus the term $[HCO_3^-]_p$ means the concentration of bicarbonate in plasma.

Total carbon dioxide of plasma exists in three forms: dissolved carbon dioxide, carbonic acid, and bicarbonate ions. Given the total carbon dioxide content of plasma and the pH, the concentration of bicarbonate and the partial pressure of carbon dioxide can be calculated.

When a gas dissolves in a liquid, the concentration of the gas in the liquid is directly proportional to the partial pressure of the gas. For carbon dioxide in plasma, this fact is represented by the equation

$$[\text{Dissolved } CO_2]_p = a' \, P_{CO_2} \tag{38}$$

where a' is the proportionality constant.

Dissolved carbon dioxide is in equilibrium with carbonic acid as expressed by the equation

$$CO_2 + H_2O \; \rightleftharpoons \; H_2CO_3. \tag{39}$$

Since the concentration of dissolved carbon dioxide is directly proportional to the P_{CO_2}, the concentration of carbonic acid must also be directly proportional to it. If the dissolved carbon dioxide and the carbonic acid concentrations are each proportional to the P_{CO_2}, their sum also is. This fact can be expressed by the equation

$$[\text{Dissolved } CO_2 + H_2CO_3]_p = a \, P_{CO_2}. \tag{40}$$

The equilibrium represented by equation (39) is far to the left, and in plasma the concentration of dissolved carbon dioxide is about 1,000 times higher than the concentration of carbonic acid. For the sake of brevity, the term $[\text{Dissolved } CO_2 + H_2CO_3]$ can be written as $[CO_2]$, and this term is understood to mean the sum of the concentrations of dissolved carbon dioxide and carbonic acid, of which the concentration of dissolved carbon dioxide is by far the larger part. In subsequent equations the symbol $[CO_2]$ will have this inclusive meaning.

Using this definition, equation (40) can be written

$$[CO_2]_p = a \, P_{CO_2}. \tag{41}$$

The proportionality constant, a, in this equation is only slightly different numerically from the constant a' in equation (38).

Carbonic acid ionizes according to the equation

$$H_2CO_3 \; \rightleftharpoons \; HCO_3^- + H^+. \tag{42}$$

It can be shown theoretically and confirmed experimentally that the relation between substances represented in this equation can be expressed by the mass-action law: the product of the concentrations of substances on the right divided by the concentration of the substance on the left is equal to a constant.

$$\frac{[H^+][HCO_3^-]}{[H_2CO_3]} = K'. \tag{43}$$

The concentration of carbonic acid is proportional to the concentration of dissolved carbon dioxide. Consequently, the term $[CO_2]$ can be substituted for the term $[H_2CO_3]$ in the denominator, and the numerical value of the constant is changed. This gives the equation

$$\frac{[H^+][HCO_3^-]}{[CO_2]} = K. \tag{44}$$

Taking the logarithm of both sides of the equation, we have

$$\log \frac{[H^+][HCO_3^-]}{[CO_2]} = \log K. \tag{45}$$

The logarithm of the product of two quantities is equal to the sum of the logarithms of the quantities, and

$$\log [H^+] + \log \frac{[HCO_3^-]}{[CO_2]} = \log K. \tag{46}$$

Transposing, we have

$$\log [H^+] = \log K - \log \frac{[HCO_3^-]}{[CO_2]}. \tag{47}$$

Changing signs on both sides, we have

$$-\log [H^+] = -\log K + \log \frac{[HCO_3^-]}{[CO_2]}. \tag{48}$$

Since $-\log [H^+]$ is pH, and $-\log K$ is called pK:

$$pH = pK + \log \frac{[HCO_3^-]}{[CO_2]}. \tag{49}$$

The pH of the plasma can be measured, but there are no direct analytical methods for measuring the bicarbonate and dissolved carbon dioxide concentrations. The two quantities which can be measured are the total carbon dioxide concentration and the P_{CO_2}. According to equation (41), the concentration of dissolved carbon dioxide is directly proportional to the P_{CO_2}. Therefore, the term $(a\,P_{CO_2})$ can be substituted in the denominator of equation (49) giving

$$pH = pK + \log \frac{[HCO_3^-]}{a\,P_{CO_2}}. \tag{50}$$

Total carbon dioxide of plasma is the sum of bicarbonate and dissolved carbon dioxide concentrations. If the total carbon dioxide concentration is known, the bicarbonate concentration can be calculated by subtraction:

$$[\text{Total } CO_2]_p = [CO_2]_p + [HCO_3^-]_p, \tag{51}$$

$$[HCO_3^-]_p = [\text{Total } CO_2]_p - [CO_2]_p. \tag{52}$$

Substituting equation (41) in the last equation, we have

$$[HCO_3^-]_p = [\text{Total } CO_2]_p - a\,P_{CO_2}. \tag{53}$$

Substituting the right-hand term in equation (50), we have

$$pH = pK + \log \frac{[\text{Total } CO_2]_p - a\,P_{CO_2}}{a\,P_{CO_2}}. \tag{54}$$

This equation contains two constants, each of which has been measured. When the equation is applied to plasma at body temperature, when the quantity $[\text{Total } CO_2]_p$ is in *millimoles per liter*, and when the P_{CO_2} is in mm Hg

$$pK = 6.10* \qquad \text{and} \qquad a = 0.0301.$$

Therefore,

$$pH = 6.10 + \log \frac{[\text{Total } CO_2]_p - 0.0301\,P_{CO_2}}{0.0301\,P_{CO_2}}. \tag{55}$$

The final equation contains three unknowns: pH, [Total $CO_2]_p$ and P_{CO_2}. If any two are measured, the third can be calculated.

1.14. Calculation of the Partition of Carbon Dioxide in Plasma

When equation (55) is used, the term $[\text{Total } CO_2]_p$ must be in millimoles per liter. Analytical figures for this quantity are sometimes reported in volumes per cent, and the figures must be converted to millimoles per liter. Volumes per cent is defined as the number of cubic centimeters of gas contained in 100 milliliters of fluid. If the total carbon dioxide is x volumes per cent, there are x cc of gas per 100 milliliters or $10x$ cc per liter. One mole of carbon dioxide at standard temperature and pressure occupies

---

*The value of pK depends upon the method by which it is determined, upon the temperature, and upon the pH. Except for the most precise work, differences from 6.10 can be ignored. See Siggaard-Andersen, 1962, *Scand. J. Clin. Lab. Invest.* 14:587.

22.26 liters, or 22,260 cc. (Note that the volume occupied by carbon dioxide under these conditions is slightly smaller than the 22.4 liters occupied by other gases.) Hence

$$10x \text{ cc per liter} = \frac{10x}{22,260} \text{ moles per liter,} \qquad (56)$$

$$\frac{10x}{22,260} \text{ moles per liter} = \frac{1,000(10x)}{22,260} \text{ millimoles per liter,} \qquad (57)$$

$$\text{Millimoles per liter} = \frac{x \text{ vol } \%}{2.226}. \qquad (58)$$

The number of volumes per cent divided by 2.226 equals the number of millimoles per liter.

The method of using equation (55) to calculate the partition of carbon dioxide in plasma is illustrated by example 11.

Example 11.   Plasma samples of arterial and venous blood were analyzed for pH and total carbon dioxide. The observed values are given in the first two lines of table 2. Calculate the P_{CO_2}, $[CO_2]_p$, and $[HCO_3^-]_p$.

Table 2

	Arterial	Venous
pH, by measurement	7.44	7.39
Total CO_2, vol %, by analysis	59.4	62.0
Total CO_2, mM/l	26.7	27.8
P_{CO_2}, mm Hg	39	45
$[CO_2]_p$, mM/l	1.2	1.4
$[HCO_3^-]_p$, mM/l	25.5	26.4

First Step.   The total carbon dioxide in volumes per cent must be converted to millimoles per liter.

$(59.4 \text{ vol } \%)/2.226 = 26.7 \text{ mM/l,}$
$(62.0 \text{ vol } \%)/2.226 = 27.8 \text{ mM/l.}$

Second Step.   Substitute the known values in equation (55). For the arterial sample the equation becomes

$$7.44 = 6.10 + \log \frac{26.7 - 0.0301\, P_{CO_2}}{0.0301\, P_{CO_2}},$$

$$1.34 = \log \frac{26.7 - 0.0301\, P_{CO_2}}{0.0301\, P_{CO_2}}.$$

Take the antilogarithm of both sides of the equation:

$$\text{antilog } 1.34 = \frac{26.7 - 0.0301\, P_{CO_2}}{0.0301\, P_{CO_2}}.$$

The antilogarithm of 1.34 is 21.88:

$$21.88 = \frac{26.7 - 0.0301\ P_{CO_2}}{0.0301\ P_{CO_2}},$$

$$21.88\,(0.0301\ P_{CO_2}) = 26.7 - 0.0301\ P_{CO_2},$$

$$0.658\ P_{CO_2} + 0.0301\ P_{CO_2} = 26.7,$$

$$0.688\ P_{CO_2} = 26.7,$$

$$P_{CO_2} = 26.7/0.688 = 39 \text{ mm Hg}.$$

Third Step. The concentration of dissolved carbon dioxide is given by the equation

$$[CO_2]_p = 0.0301\ P_{CO_2}.$$

Substituting the calculated value of P_{CO_2}:

$$[CO_2]_p = 0.0301\ (39)$$
$$= 1.2 \text{ mM/l}.$$

Fourth Step. The bicarbonate concentration is the difference between total carbon dioxide and dissolved carbon dioxide:

$$[HCO_3^-]_p = 26.7 - 1.2 = 25.5 \text{ mM/l}.$$

Fifth Step. Repeat the calculations for venous blood. The values obtained are entered in table 2.

Calculations using equation (55) are standard mathematical operations. If the values are assumed for any two variables, the necessary third variable can be calculated. Thus, if one assumes that the pH is 7.40 and the total carbon dioxide concentration is 30 millimoles per liter, the P_{CO_2} must be 47 mm Hg. One could continue in this way to assume values for all possible combinations and to calculate the necessary value for the third variable. Once this is done, the calculations need never be repeated.

The results obtained by this series of calculations could be entered in a table, but the table would be long and difficult to use. Instead, the values can be expressed in a nomogram. A nomogram is a chart on which lines are drawn whose length and shape have values satisfying an equation or an empirical relationship. A straight line placed across the nomogram cuts the lines at values satisfying the equation. A nomogram describing the carbon dioxide system of plasma is shown in figure 14.

To use the nomogram, place a straight edge, preferably a transparent ruler, across the lines so that it cuts the left hand one at the known value of total carbon dioxide concentration of the plasma and cuts the pH line at the known pH. Then read the values of P_{CO_2}, dissolved carbon dioxide, and bicarbonate where the straight edge cuts the other lines.

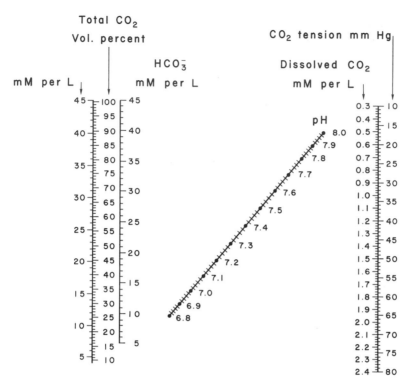

Fig. 14 Nomogram of equation (45). From McLean, 1938, *Physiol. Rev.* 18:495. (Reproduced by permission.)

1.15. Calculation of the Partition of Carbon Dioxide in Whole Blood

Whole blood is a heterogeneous system composed of plasma and erythrocytes. Although the P_{CO_2} is the same within plasma and erythrocytes, the concentrations of dissolved carbon dioxide and bicarbonate in the two phases differ. The reason will be explained below. Concentration of these within erythrocytes can be calculated from analytical data obtained on whole blood and plasma.

Total carbon dioxide in a liter of blood is equal to the sum of carbon dioxide in the plasma of that liter plus carbon dioxide in the erythrocytes of that liter. The fraction of the volume of a liter of blood which is made up of erythrocytes is called V_c. Likewise, the fraction of a liter of blood which is made up of plasma is V_p, and V_p equals $(1 - V_c)$. The expression [Total $CO_2]_c$ means the number of millimoles of carbon dioxide in one liter of erythrocytes. Since there are V_c liters of erythrocytes in one liter of blood, there are V_c [Total CO$_2$]$_c$ millimoles of carbon dioxide in the erythrocytes of one liter of blood:

$$\text{CO}_2 \text{ in cells of one liter} = V_c \text{ [Total CO}_2]_c. \qquad (59)$$

Likewise, the expression [Total CO$_2$]$_p$ means the number of millimoles of carbon dioxide in one liter of plasma. Since there are V_p liters of plasma in one liter of blood, there are V_p [Total CO$_2$]$_p$ millimoles of carbon dioxide in the plasma of one liter of blood:

$$\text{CO}_2 \text{ in plasma of 1 liter} = V_p \text{ [Total CO}_2]_p. \qquad (60)$$

The expression [Total CO$_2$]$_b$ means the number of millimoles of carbon dioxide in one liter of whole blood. This is the sum of the carbon dioxide in the erythrocytes and plasma. Therefore,

$$[\text{Total CO}_2]_b = V_c \text{ [Total CO}_2]_c + V_p \text{ [Total CO}_2]_p, \qquad (61)$$

or

$$[\text{Total CO}_2]_b = V_c \text{ [Total CO}_2]_c + (1 - V_c)$$
$$[\text{Total CO}_2]_p. \qquad (62)$$

This equation gives

$$[\text{Total CO}_2]_c = \frac{[\text{Total CO}_2]_b - (1 - V_c) \, [\text{Total CO}_2]_p}{V_c}. \qquad (63)$$

Since V_c, [Total CO$_2$]$_b$ and [Total CO$_2$]$_p$ can be measured, the concentration of carbon dioxide in erythrocytes can be calculated.

Partition of total carbon dioxide of erythrocytes into its various forms is more complicated than the partition in plasma. Carbon dioxide is carried in erythrocytes in these forms: dissolved carbon dioxide, carbonic acid, bicarbonate ions, and carbamino—CO$_2$. As in plasma, no distinction is made between dissolved carbon dioxide and carbonic acid, and the sum is called "dissolved carbon dioxide."

If the P_{CO_2} of plasma is known either by direct measurement or by calculation, the P_{CO_2} of erythrocytes is known, for the two are the same. The concentration of dissolved carbon dioxide in erythrocytes can be calculated from the P_{CO_2} and the solubility coefficient whose value for erythrocytes is 0.025 when the P_{CO_2} is in mm Hg and [CO$_2$]$_c$ is in millimoles per liter.

$$[\text{CO}_2]_c = 0.025 \, P_{\text{CO}_2}. \qquad (64)$$

If the total carbon dioxide of erythrocytes is calculated by means of equation (63) and the dissolved carbon dioxide is calculated by means of equation (64), the difference between the two is carbon dioxide carried as bicarbonate ions and carb-amino—CO_2.

$$[\text{Total } CO_2]_c - [CO_2]_c = [HCO_3^-]_c + [\text{Carbamino}—CO_2]_c. \quad (65)$$

The amount of carbamino—CO_2 can be calculated on the basis of several assumptions, but except for research purposes, this calculation is not particularly useful. Consequently, it is usually omitted, and the difference between total carbon dioxide and dissolved carbon dioxide is called the "bicarbonate concentration of erythrocytes." This is not strictly true, as the data in table 1 show. However, as far as the buffering processes of blood are concerned, the distinction between carbamino—CO_2 and bicarbonate can be ignored. If the two forms of carbon dioxide are called "bicarbonate," equation (65) must be rewritten as

$$[\text{Total } CO_2]_c - [CO_2]_c = \text{"}[HCO_3^-]_c\text{."} \quad (66)$$

This equation will be used in the discussion to follow, but it is not strictly true. The assumption underlying the equation cannot be used when diffusion relations between plasma and erythrocytes are being considered, for bicarbonate ions are diffusable, while carbamino—CO_2, being attached to hemoglobin molecules, is not.

Example 12. The same samples of arterial and venous blood described in example 11 were found to have total carbon dioxide in whole blood and V_c as shown in the first two lines of table 3.

Table 3

	Arterial	Venous
Total CO_2, whole blood, vol %, by analysis	45.8	47.5
V_c, by measurement	0.47	0.51
P_{CO_2}, mm Hg, by calculation, example 11	39	45
V_p	0.53	0.49
Total CO_2, whole blood, mM/l	20.6	21.3
Total CO_2, plasma, mM/l, example 11	26.7	27.8
Total CO_2, erythrocytes, mM/l	13.7	15.0
$[CO_2]_c$, mM/l	1.0	1.1
"$[HCO_3^-]_c$," mM/l	12.7	13.9

Calculate the concentrations of dissolved carbon dioxide and bicarbonate in the erythrocytes.

First Step. In the arterial sample,

$$V_p = (1 - V_c) = (1 - 0.47) = 0.53.$$

In the venous sample,
$$V_p = (1 - V_c) = (1 - 0.51) = 0.49.$$

Second Step.   In the arterial sample,
$$(45.8 \text{ vol } \%)/2.226 = 20.6 \text{ mM/l}.$$

In the venous sample,
$$(47.5 \text{ vol } \%)/2.226 = 21.3 \text{ mM/l}.$$

Third Step.   In the arterial sample by equation (63),
$$[\text{Total } CO_2]_c = \frac{20.6 - 0.53(26.7)}{(0.47)} = 13.7 \text{ mM/l}.$$

In the venous sample,
$$[\text{Total } CO_2]_c = \frac{21.3 - 0.49(27.8)}{(0.51)} = 15.0 \text{ mM/l}.$$

Fourth Step.   In the arterial sample by equation (64),
$$[CO_2]_c = 0.025(39) = 1.0 \text{ mM/l}.$$

In the venous sample,
$$[CO_2]_c = 0.025(45) = 1.1 \text{ mM/l}.$$

Fifth Step.   In the arterial sample by equation (66),
$$\text{``}[HCO_3^-]_c\text{''} = (13.7 - 1.0) = 12.7 \text{ mM/l}.$$

In the venous sample,
$$\text{``}[HCO_3^-]_c\text{''} = (15.0 - 1.1) = 13.9 \text{ mM/l}.$$

1.16. The pH-Bicarbonate Diagram

The fundamental data useful in the study of the acid-base pattern of plasma are the pH, the P_{CO_2}, and the bicarbonate concentration. These are related by the equation

$$\text{pH} = 6.10 + \log \frac{[HCO_3^-]_p}{0.0301 \, P_{CO_2}}. \tag{67}$$

A useful way of plotting these variables is the pH-bicarbonate diagram shown in figure 15. The pH units are plotted as abscissas, and the bicarbonate concentrations are plotted as ordinates.

At any point on the graph where a line representing pH crosses a line representing bicarbonate concentration, there is a unique value of P_{CO_2}. If pH and bicarbonate concentrations are known, P_{CO_2} must have a single value which will satisfy equation (67). On the other hand, if pH and P_{CO_2} are known, there can be only one value of bicarbonate concentration which will satisfy the equation. Therefore, we can find the locus of all points on the graph at which the P_{CO_2} must be 40 mm Hg. All points lying on the locus have the same pressure of carbon dioxide, and the locus is the P_{CO_2} *isobar*.

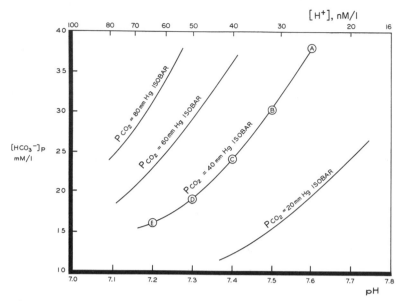

Fig. 15. The pH-bicarbonate diagram with P_{CO_2} isobars for P_{CO_2}'s equal to 20, 40, 60, and 80 mm Hg. The concentration of hydrogen ions [H$^+$] in nanomoles per liter is given by the scale at the top of the figure.

Example 13. Determine and plot the P_{CO_2} isobar for P_{CO_2} equals 40 mm Hg.

First Step. Assume that the pH is 7.60. Calculate the bicarbonate concentration.

$$7.60 = 6.10 + \log \frac{[HCO_3^-]_p}{0.0301(40)},$$

$$1.50 = \log \frac{[HCO_3^-]_p}{1.2}.$$

The antilogarithm of 1.50 is 31.6; hence

$$31.6 = [HCO_3^-]_p/1.2,$$
$$[HCO_3^-]_p = 37.9 \text{ mM/l}.$$

When P_{CO_2} is 40 mm Hg and pH is 7.60, the bicarbonate concentration must be 37.9 millimoles per liter. This is plotted as point A in figure 15.

Second Step. When these calculations are repeated for each of the pairs of values for pH and P_{CO_2} given below, the bicarbonate concentrations given in the third column are found.

P_{CO_2}, mm Hg	pH	$[HCO_3^-]_p$, mM/l
40	7.50	30.1
40	7.40	24.0
40	7.30	19.0
40	7.20	15.1
40	7.10	12.0

These points are plotted as *B, C, D, E,* and *F.* When the points are connected by a smooth curve, the P_{CO_2} equals 40 mm Hg isobar is obtained. Any pair of values of pH and bicarbonate concentration on plasma whose P_{CO_2} is 40 mm Hg must fall on this line. Isobars for other P_{CO_2}'s can be calculated, and those for 20, 60, and 80 mm Hg are plotted on the figure.

1.17. The Buffer Value of Separated Plasma and of Oxygenated Whole Blood

When the acid-base pattern of blood is studied by methods described here, the data used are the pH, P_{CO_2}, and bicarbonate concentration of plasma. Section 1.11 explains that the major part of the buffering power of blood is in erythrocytes and that plasma alone is a poor buffer. For this reason a distinction is made between *true plasma* and *separated plasma.*

True plasma is plasma removed from erythrocytes under rigid anaerobic conditions so that no carbon dioxide is lost before the plasma is analyzed. If any change in the P_{CO_2} is made for the purpose of studying the buffering power of blood, the change is made before the plasma is removed for analysis. Therefore, the plasma pH and the bicarbonate concentration are affected by the buffering power of the erythrocytes. When the data are plotted on a pH-bicarbonate diagram, they show the buffering power of plasma plus erythrocytes.

Separated plasma is plasma removed from erythrocytes before any change is made in the P_{CO_2}. The pH and bicarbonate concentration of separated plasma do not reflect the buffering power of hemoglobin, but any changes observed are results of the buffering power of plasma alone.

Example 14.   A sample of blood was obtained, and the plasma was separated from the erythrocytes. Four small samples were equilibrated with gas mixtures containing various percentages of carbon dioxide. The temperature was 37°C, and the barometric pressure was 655 mm Hg. At the end of equilibration the gas phases and the plasma were analyzed for carbon dioxide. The results are entered in the first two lines of table 4. Calculate the P_{CO_2}, pH, and bicarbonate concentration of each sample, and plot the results on a pH-bicarbonate diagram.

First Step. Calculate the P_{CO_2}. Correction must be made for the vapor pressure of water. The dry gas pressure was $(655 - 47 \text{ mm Hg})$, or 608 mm Hg. For the first sample

Table 4

| | Separated Plasma No. | | | |
	1	*2*	*3*	*4*
% CO_2 in gas phase	10.9	8.3	5.3	3.9
Total CO_2, plasma, mM/l	27.0	26.0	24.5	23.7
P_{CO_2}, mm Hg	66.2	50.5	32.2	23.7
$[CO_2]_p$, mM/l	2.0	1.5	1.0	0.7
$[HCO_3^-]_p$, mM/l	25.0	24.5	23.5	23.0
pH	7.20	7.30	7.48	7.60

$$P_{CO_2} = 608(10.9)/100 = 66.2 \text{ mm Hg.}$$

Second Step. Calculate the dissolved carbon dioxide concentration.

$$[CO_2]_p = 0.0301(66.2) = 2.0 \text{ mM/l.}$$

Third Step. Calculate the bicarbonate concentration, which is the difference between the total carbon dioxide and the dissolved carbon dioxide concentrations.

$$[HCO_3^-]_p = 27.0 - 2.0 = 25.0 \text{ mM/l.}$$

Fourth Step. Calculate the pH.

$$pH = 6.10 + \log \frac{25.0}{2.0},$$
$$= 6.10 + \log 12.5.$$

The logarithm of 12.5 is 1.10. Therefore,

$$pH = 6.10 + 1.10 = 7.20.$$

The calculated values for all samples are entered in table 4.

The points are plotted as open circles in figure 16, and the straight line labeled "Separated Plasma" is drawn through the points.

When separated plasma is equilibrated with gas mixtures having a high P_{CO_2} it is titrated with carbonic acid. Carbon dioxide dissolves in the plasma and forms carbonic acid which in

turn gives hydrogen ions and bicarbonate. The increase in bicarbonate is equal to the amount of acid added to the plasma,
and it is therefore a measure of the amount of acid added. Some
hydrogen ions combine with buffers of the plasma and disappear.
The rest of the hydrogen ions increase the hydrogen ion concentration of the plasma and reduce its pH.

When the P_{CO_2} is reduced to low values, carbon dioxide is
removed from the plasma. This carbon dioxide is formed from
bicarbonate and hydrogen ions. The decrease in bicarbonate
concentration is equal to the amount of acid removed from the
plasma, and it is therefore a measure of the amount removed.
Some of the hydrogen ions are given up by buffers of the plasma,
and the rest come from free hydrogen ions in solution in the
plasma, thereby increasing its pH.

The buffer value of a solution is the amount of acid which
must be added to cause a change of one pH unit. When separated
plasma is titrated with carbon dioxide, the amount of carbonic
acid added can be measured by the change in bicarbonate concentration. By dividing the change in bicarbonate concentration
by the change in pH, the buffer value is determined.

$$\text{Buffer value} = \Delta HCO_3^- / \Delta pH. \tag{68}$$

Figure 14 shows that the bicarbonate concentration falls by 4.3
millimoles per liter as the pH rises 0.8 units.

Buffer value $= -4.3/0.8$
$= -5.4$ mM/l per pH unit.

The line in figure 16 representing the buffer curve of plasma
is flat; plasma is a poor buffer. On the other hand, whole blood
is a good buffer; and this fact is illustrated by the following data
on true plasma.

Example 15.   Four samples of whole blood were equilibrated at 37°C with gas mixtures containing various percentages of carbon dioxide. The barometric pressure was
758 mm Hg. Plasma was removed from the erythrocytes
under anaerobic conditions, and true plasma samples were
analyzed for carbon dioxide. In each instance the P_{O_2} of
the gas phase was high enough to ensure complete saturation of the hemoglobin.

The percentages of carbon dioxide in the gas phases
and the total carbon dioxide concentrations of the true
plasma samples are entered in the first two lines of table 5.
The P_{CO_2}, the pH, and the bicarbonate concentrations were
calculated by the methods outlined in example 14.

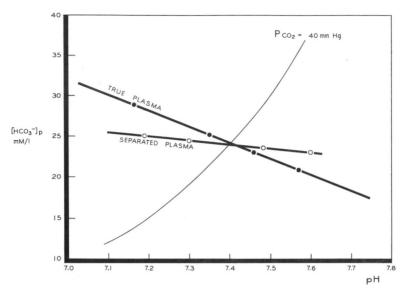

Fig. 16.   Buffer curves of separated plasma and oxygenated true plasma determined *in vitro*. Data for oxygenated true plasma from the blood of A.V.B. described in L. J. Henderson, *Blood* (Yale University Press: New Haven, 1928). (Reproduced by permission.)

The points describing the buffering of carbon dioxide by the blood of the A.V.B. are plotted as filled circles in figure 16, and the straight line labeled "True Plasma" is drawn through them.

The buffer line of true plasma is much steeper than that of separated plasma. Its slope or buffer value, calculated by means of equation (68), is -21.6 millimoles per liter per pH unit. The reason is that the buffer line of true plasma reflects the buffering

Table 5

	True Plasma No.			
	1	*2*	*3*	*4*
% CO_2, gas phase	11.97	6.55	4.69	3.28
Total CO_2, plasma, mM/l	32.3	26.4	24.0	21.5
P_{CO_2}, mm Hg	85.1	46.5	33.3	23.3
$[CO_2]_p$, mM/l	2.6	1.4	1.0	0.7
$[HCO_3^-]_p$, mM/l	29.7	25.0	23.0	20.8
pH	7.16	7.35	7.46	7.57

SOURCE: Data on the blood of A.V.B. adapted from L. J. Henderson, *Blood* (Yale University Press: New Haven, 1928). (Reproduced by permission.)

power of hemoglobin. When carbon dioxide is added to whole blood at a high P_{CO_2}, its hydrogen ions are buffered by hemoglobin, and most of the newly formed bicarbonate diffuses into the plasma. Therefore, the increase in plasma bicarbonate reflects the effect of hemoglobin as a buffer. Conversely, when carbon dioxide is removed from whole blood at a low P_{CO_2}, bicarbonate ions diffuse from plasma into erythrocytes where they combine with hydrogen ions given up by hemoglobin.

The fact that the units of buffer value are millimoles per liter and pH units obscures the real magnitude of the buffering process.

The difference between the bicarbonate concentrations of the true plasma samples 1 and 4 in table 5 is (29.7 − 20.8) or 8.9 millimoles per liter. The hydrogen ion concentration of sample 1, calculated from pH 7.16, is 69 nanomoles, or 0.000,069 millimoles per liter. The hydrogen ion concentration of sample 4 is 0.000,027 millimoles per liter, and the difference between the two samples is 0.000,042 millimoles per liter. The change in bicarbonate concentration is 211,900 times greater than the change in hydrogen ion concentration.

What Happens in a Person 2

2.1. The Slope of the Normal Buffer Line
In Vitro and In Vivo

The buffer value of blood *in vitro* expressed by the slope of the line relating bicarbonate concentration of true plasma to pH depends upon the concentration of hemoglobin in whole blood. The data in table 6 give the bicarbonate concentrations of true plasma at three pH values for blood containing 5, 10, 15, and 20 grams of hemoglobin per 100 milliliters. The points are plotted and connected by straight lines in figure 17.

Table 6

Hb, g %	5	10	15	20
Hb, mEq/l	3.0	6.0	9.0	12.0
$[HCO_3^-]_p$ at pH 7.30, mM/l	25.9	26.6	27.3	28.0
$[HCO_3^-]_p$ at pH 7.40, mM/l	24.4	24.4	24.4	24.4
$[HCO_3^-]_p$ at pH 7.50, mM/l	22.9	22.2	21.5	20.8

All data described so far were obtained on blood *in vitro*. Blood was drawn from the subject, alterations in its P_{CO_2} were effected, and the bicarbonate concentration of plasma and the pH were determined. There are two major reasons why the slope of the buffer line of blood *in vivo* might differ from that of the same blood *in vitro*.

1. Electrolytes of plasma are in equilibrium with electrolytes of interstitial fluid, and any change in the bicarbonate concentration of plasma is rapidly followed by redistribution of bicarbonate ions between interstitial fluid and plasma which restores equilibrium. If the P_{CO_2} of plasma *in vivo* is increased, hydrogen ions and bicarbonate ions are formed, and these distribute themselves between blood and interstitial fluid. Because the protein content of interstitial fluid is, on the average, lower than that of plasma, interstitial fluid is an even poorer buffer than plasma. Therefore, hydrogen ions are buffered chiefly by hemoglobin of erythrocytes, and the fall in pH is governed by the concentration of hemoglobin in blood. On the other hand, bicarbonate ions distribute themselves between plasma and interstitial fluid in such a way that their concentrations in each

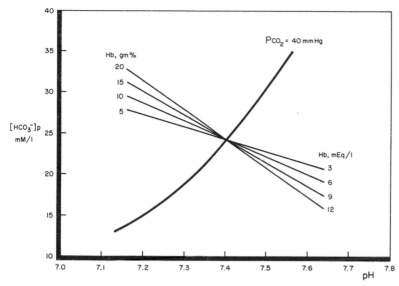

Fig. 17.   Buffer curves of true plasma of blood containing 5, 10, 15, or 20 grams of hemoglobin per 100 ml. Hemoglobin concentrations in milliequivalents per liter are also given.

fluid are almost identical. Since the volume of interstitial fluid is usually two to three times greater than the plasma volume, only one-third to one-fourth of the increment in bicarbonate ions remains in the plasma; the rest goes to the interstitial fluid. For this reason the buffer line of the system, blood plus interstitial fluid, is flatter than that of blood alone. For any rise in P_{CO_2}, the rise in plasma bicarbonate concentration is less for this system than it is for whole blood alone.

2. All cells of the body, not only erythrocytes, contain buffers. When the body's P_{CO_2} rises, all cells buffer hydrogen ions, and for each hydrogen ion buffered a bicarbonate ion is formed. Most of these bicarbonate ions accumulate in interstitial fluid and plasma. Consequently, there is a rise in plasma bicarbonate concentration which results, not from blood buffering, but from whole body buffering. Therefore, there is a rise in plasma bicarbonate concentration without a corresponding fall in pH.

These two *in vivo* processes, one reducing the change in plasma bicarbonate concentration below that to be expected in blood alone and the other rising it above that to be expected, affect the slope of the buffer line of blood in opposite ways. The actual resulting *in vivo* buffering line cannot be exactly predicted; it must be measured.

For determination of the carbon dioxide titration curve *in vivo,* normal men were placed in a chamber in which they breathed gas mixtures containing oxygen at a P_{O_2} of 140 to 160 mm Hg and carbon dioxide at a P_{CO_2} of zero to 80 mm Hg. Arterial blood samples were taken for analysis beginning 10 minutes after entering the chamber and at other times up to 60 minutes. The acid-base status of each subject was found to be stable throughout this period. The data for the seven subjects are plotted as open circles at the left of the pH-bicarbonate diagram in figure 18. The heavy curved lines surrounding the points are the limits which contain, with estimated 95% probability, anticipated responses of normal human subjects during uncomplicated respiratory acidosis. Only at the highest P_{CO_2} is there a systematic deviation from the *in vitro* buffer line of A.V.B.'s blood.

Twelve anesthetized patients undergoing elective operations were subjected to artificial hyperventilation as an adjunct to general anesthesia. The acid base status of their arterial blood samples is plotted as filled circles at the right of the pH-bicarbonate diagram in figure 18. The heavy curved lines sur-

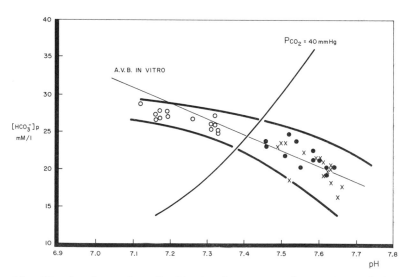

Fig. 18. *In vivo* carbon dioxide titration curves of men. Open circles are for men breathing gas mixtures having high P_{CO_2}'s, and the heavy lines are the 95% confidence limits derived from the data. Filled circles are for men being hyperventilated, and the heavy lines are the 95% confidence limits derived from the data. Crosses are for men voluntarily hyperventilating. Data from Arbus, Hebert, Levesque, Etsten, and Schwartz, 1969, *New Eng. J. Med.* 280:117; Brackett, Cohen, and Schwartz, 1965, *New Eng. J. Med.* 272:6; Eldridge and Salzer, 1967, *J. Appl. Physiol.* 22:401; and Elkinton, Singer, Barker, and Clark, 1955, *J. Clin. Invest.* 34:1671. (Reproduced by permission.)

rounding the points are the limits calculated from these data to contain, with 95% probability, anticipated responses of normal subjects during uncomplicated respiratory alkalosis.

In the same figure data obtained on normal alert men undergoing voluntary hyperventilation are plotted as crosses. Their acid base status did not differ from that of the anesthetized patients during hyperventilation.

These data show that the relation between plasma pH and bicarbonate concentration in normal men *in vivo* is curvilinear rather than rectilinear. The equation for a curve fitting the *in vivo* data is

$$[HCO_3^-]_p = 31.39 \frac{P_{CO_2}}{P_{CO_2} + 12.95}. \tag{69}$$

The coordinates of this curve are given in table 7. This *whole-body buffer line* together with the 95% confidence limits is plotted in figure 19.

The ability of the body to buffer acid produced by carbon dioxide can be expressed in terms of hydrogen ion concentration as well as in terms of pH. For the data on acute hypercapnia shown in figure 18 the change in $[H^+]_p$ for a change in P_{CO_2}, $\Delta[H^+]_p/\Delta P_{CO_2}$, is 0.77 nanomoles per mm Hg. For the data on acute hypocapnia $\Delta[H^+]_p/\Delta P_{CO_2}$ is 0.74 nanomoles per mm Hg. The value for $\Delta[H^+]_p/\Delta P_{CO_2}$ implied by equation (69) is 0.76 nanomoles per mm Hg.

The data show that the whole-body buffer line determined *in vivo* is not quite the same as the buffer line of blood *in vitro*. In later sections of this book methods of calculating the magnitude of deviations of acid-base status from normality are described. In the calculation of some of these—base excess, base

Table 7

Coordinates of the Human Whole-Body Buffer Line

Plasma P_{CO_2}	Plasma pH	$[HCO_3^-]_p$
20 mm Hg	7.60	19.1 mM
30	7.49	21.9
40	7.40	23.7
50	7.32	24.9
60	7.25	25.8
80	7.15	27.0

SOURCE: Computed from equation (69), which was derived by Dr. Josef R. Smith from the data of Brackett, Cohen, and Schwartz, 1965, *New Eng. J. Med.* 272:6, and Arbus, Hebert, Levesque, Etsten, and Schwartz, 1969, *New Eng. J. Med.* 280:117. (Reproduced by permission.)

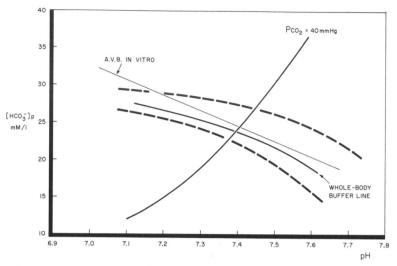

Fig. 19. Whole-body buffer line determined *in vivo* compared with the *in vitro* buffer line of A.V.B.'s blood. Whole-body buffer line was plotted from equation (69) derived by J. R. Smith from the data plotted in figure 18. The broken lines are the 95% confidence limits drawn in figure 18.

deficit, and standard bicarbonate — the slope of the buffer line *in vitro* is either assumed or is measured. Consequently, the calculations give more or less exact estimates of base excess, base deficit, and standard bicarbonate outside the body. They do not give exact estimates of the quantities in the body as a whole for the reason that the slope of the buffer line of blood in the body is not measured. That slope may or may not be similar to that of blood *in vitro*.

The clinician who wishes to use estimates of base excess or base deficit in planning his therapy may be mislead unless he realizes that the *in vivo* buffer line may be different from the *in vitro* buffer line. He may even be mislead if he assumes that the buffer line of his patient is similar to that of the normal men who served as subjects for the data plotted in figure 16. The *in vivo* slope depends in part upon the ratio of interstitial fluid volume to blood volume. Blood volume may be anything from 6 to 15 per cent of the patient's body weight. Interstitial fluid volume does not always bear a fixed relation to blood or plasma volume; it may be very low in a dehydrated patient in diabetic acidosis; or it may be very high as in a patient with congestive failure. The *in vivo* slope also depends on buffering by cells, and this in turn is a function of lean body mass. Lean body mass may form a large or small proportion of total body weight. Furthermore, body buffering may change with time. Hydrogen ions may slowly exchange for intracellular potassium ions so that the body's load of hydrogen ions may be much greater in long-standing acidosis than it is in acute hypercapnia.

There is a large difference between acute and chronic buffering of the acid derived from carbon dioxide. An example is shown in figure 20.

A dog was acutely exposed in a chamber to gas mixtures which made the P_{CO_2} of its alveolar air 40, 60, 83 and 108 mm Hg. The data on arterial blood samples obtained at the four levels of hypercapnia are plotted in figure 20 and are labeled "acute." The same dog was maintained in the chamber for six days breathing a gas mixture which kept its alveolar P_{CO_2} at 110 mm Hg. During this period renal compensation caused a rise in plasma bicarbonate concentration of 15 millimoles per liter, and the plasma pH rose from 7.02 to 7.20. Then the ambient P_{CO_2} was acutely lowered in steps while arterial blood samples were taken. The data obtained from these samples, plotted in figure 20 as "chronic," show that the buffer curve is much steeper than it was

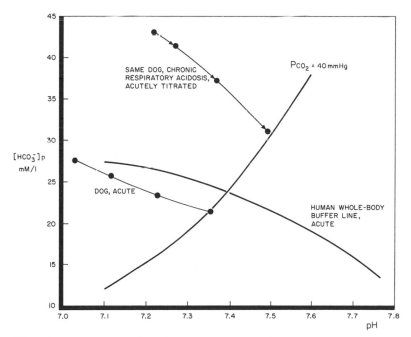

Fig. 20. Acute and chronic carbon dioxide titration curves of a dog breathing gas mixtures containing various partial pressures of carbon dioxide. Chronic curve obtained after six days of exposure to high P_{CO_2}. Data from Goldstein, Gennari, and Schwartz, 1970, *J. Clin. Invest.* 50:208. (Reproduced by permission.)

before chronic hypercapnia had occurred. The dog was almost twice as good at defending its plasma hydrogen ion concentration.

In man the acute $\Delta[H^+]_p/\Delta P_{CO_2}$ is between 0.74 and 0.77 nanomoles per mm Hg. From data obtained on patients who have been hypercapnic for some time the $\Delta[H^+]_p/\Delta P_{CO_2}$ has been calculated to be 0.24 nanomoles per mm Hg. In other words, a chronically hypercapnic person is three times better at defending his plasma hydrogen ion concentration against acid produced from carbon dioxide than is an acutely hypercapnic subject.

The chief reason for the improved ability to defend the hydrogen ion concentration in chronic hypercapnia is that the compensatory processes described in section 2.10 raise the bicarbonate concentration of the plasma. As shown in section 1.13, the relation between hydrogen ion and bicarbonate concentrations and the partial pressure of carbon dioxide is $[H^+]_p$ $[HCO_3^-]_p/P_{CO_2} = K$. This can be rearranged as $[H^+]_p = K \cdot P_{CO_2}/$ $[HCO_3^-]_p$. For a change in P_{CO_2} the corresponding change in $[H^+]_p$ is expressed as $\Delta[H^+]_p = K \cdot \Delta P_{CO_2}/[HCO_3^-]_p$. It is obvious that the larger the value of $[HCO_3^-]_p$, the smaller will be the change in hydrogen ion concentration accompanying a given change in the partial pressure of carbon dioxide.

In the following discussion of the acid-base status, the curve for A.V.B.'s blood measured *in vitro* will be used as the normal buffer line. This is done simply for convenience and for the reason that the slope of A.V.B.'s buffer line *in vitro* is not far different from that of normal men *in vivo*. However, the reader must keep two points in mind: (1) the actual buffer line of any particular person may differ from that of A.V.B. for several reasons; and (2) no sensible person would make a decision concerning the acid-base status of a patient if the validity of the decision could be vitiated by a difference between the actual slope of the buffer line of the patient's blood *in vivo* and the slope either assumed or determined *in vitro*.

2.2. The Buffer Line of Reduced Blood and the Concept of Base Excess or Deficit

Figure 12 shows that the titration curve of reduced hemoglobin is nearly parallel to the titration curve of oxyhemoglobin. This means that the buffer line of reduced blood should be nearly parallel to that of oxygenated blood as plotted in figure 16. When carbon dioxide is added to reduced blood, it forms hydrogen and bicarbonate ions in the same quantity as when it is added to oxygenated blood. Hydrogen ions titrate the reduced hemoglobin along its titration curve parallel to the titration curve of oxyhemoglobin. Conversely, removal of carbon dioxide from reduced blood removes hydrogen and bicarbonate ions and titrates reduced hemoglobin in the opposite direction.

The fact that the titration curve of reduced hemoglobin is higher than that of oxyhemoglobin means that the buffer line of reduced blood will be higher than that of oxygenated blood. Reduction of hemoglobin allows it to take up more hydrogen ions without a change in pH; and therefore, if oxygenated blood is reduced, it can take up more hydrogen ions with no change in pH. These hydrogen ions come from carbon dioxide which is added to blood at the same time oxygen is removed; and as hydrogen ions formed by ionization of carbonic acid are buffered by reduced hemoglobin, bicarbonate ions formed at the same time are distributed between erythrocytes and plasma as additional bicarbonate. At any given pH, reduced blood contains more bicarbonate than oxygenated blood, and the buffer line of reduced blood is higher than the buffer line of oxygenated blood.

Example 16.   The data given in table 8 were obtained on four samples of reduced blood of A.V.B. under the same conditions as described in example 15 with the exception that the P_{O_2} of the gas mixtures used was zero. The data are plotted in figure 21, and a straight line, labeled "Reduced True Plasma," is drawn through the points.

A buffer line of partially reduced blood would lie between the two lines plotted in figure 21.

When hemoglobin is reduced it takes up hydrogen ions. This is equivalent to adding base to the blood to remove hydrogen ions. Consider point A in figure 22, which represents the normal point of oxygenated blood. Suppose 10 millimoles of NaOH are added to a liter of the blood. This addition of base would remove hydrogen ions and cause the pH to rise. In order to restore the pH to 7.4, 10 millimoles of carbonic acid could

Table 8

	True Plasma No.			
	1	*2*	*3*	*4*
P_{CO_2}, mm Hg	83.0	54.0	33.7	22.9
$[CO_2]_p$, mM/l	2.4	1.6	1.0	0.7
$[HCO_3^-]_p$, mM/l	31.5	28.5	25.0	22.5
pH	7.20	7.34	7.49	7.61

SOURCE: Data on the reduced blood of A.V.B. adapted from L. J. Henderson, *Blood* (Yale University Press: New Haven, 1928). (Reproduced by permission.)

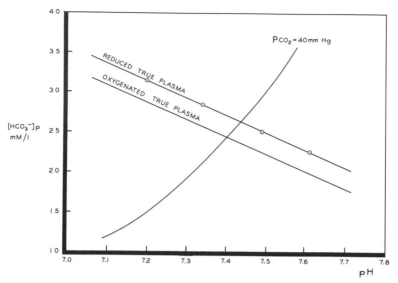

Fig. 21. Buffer curves of oxygenated true plasma and reduced true plasma. Data adapted from L. J. Henderson, *Blood* (Yale University Press: New Haven, 1928).

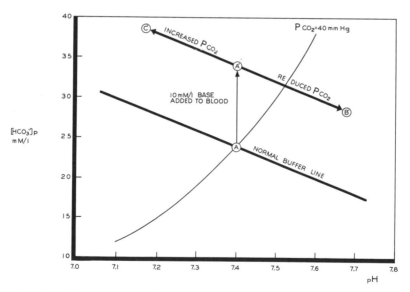

Fig. 22. The effect of adding base to blood.

be added to the blood. Hydrogen ions from carbonic acid would neutralize hydroxyl ions from the NaOH, and bicarbonate ions would increase the bicarbonate concentration by 10 millimoles per liter. The final result of addition of 10 millimoles of base and its neutralization by carbonic acid is represented by point *A'*.

The same changes occur when acid is removed from blood, when the kidneys secrete acid urine, or when hydrochloric acid is lost from the stomach by vomiting. Removal of 10 millimoles of acid per liter of blood by these means also brings the blood to point A'. Such a change occurring either by addition of base or by removal of acid is the condition of *base excess*.

Base excess is measured by titration of a blood sample with strong acid (HCl or its equivalent) to pH 7.40 at a P_{CO_2} of 40 mm Hg and at 37°C.* Such a titration of blood represented by point A' would return the blood to point A, and 10 millimoles of strong acid would be consumed in the process of neutralizing the excess base originally added to the blood.

After base excess has accumulated, the buffers of the blood are the same as they were originally, and the buffering power of the blood is unchanged. Therefore, the buffer line of blood to which excess base has been added should be parallel to its buffer line before the base was added. If the P_{CO_2} is reduced, carbonic acid is lost, the pH increases, and the bicarbonate concentration falls. The blood moves toward point B' in figure 22. If the P_{CO_2} is increased, carbonic acid is added, the pH decreases, and the bicarbonate concentration rises. The blood moves toward point C'.

Addition of acid to blood results in changes opposite those occurring when excess base is added. If 10 millimoles of hydrochloric acid are added to a liter of blood, the pH of the blood is reduced. In order to restore the pH to 7.40, 10 millimoles of carbonic acid can be removed, and this reduces the bicarbonate concentration by 10 millimoles per liter. The point representing blood to which acid has been added lies below the normal buffer line.

Acid can be added to blood by accumulation of ketone acids as in diabetes mellitus or in starvation. Removal of base from blood is equivalent to addition of acid, and this occurs as the result of loss of alkaline fluids from the digestive tract. Because addition of acid and removal of base are equivalent, the state produced by either one is called *base deficit* or *negative base excess*.

Base deficit is measured by titration of a blood sample with NaOH to pH 7.40 at a P_{CO_2} of 40 mm Hg and at 37°C.

2.3. Estimation of Base Excess or Deficit

Base excess or deficit is *not* simply the difference between the bicarbonate concentration found in a particular sample of blood

---

*This operational definition of base excess and the similar one for base deficit are from Siggaard-Andersen, 1963, *Scand. J. Clin. Lab. Invest.*, vol. 15, suppl. 70.

and the normal bicarbonate concentration in blood having no base excess or deficit. The reason is that in addition to changes effected by base excess or deficit, bicarbonate concentration is affected by respiratory adjustments.

Consider the blood represented by point *B* in figure 23. The pH is 7.50, and the plasma bicarbonate concentration is 32.5 millimoles per liter. The difference between the bicarbonate concentration of that sample and the normal value represented by point *A* is measured by arrow *1,* and it is 8 millimoles per liter. This is not the total base excess, for point *B* is displaced to the right of point *A.* This means that the pH of the blood has increased and buffers of the blood have given up some hydrogen ions to neutralize excess base. In order to return the blood to pH 7.40, more hydrogen ions must be added. This can be done by increasing the P_{CO_2}, in which event the blood will move along its buffer line to point *A'* as shown by arrow *3.* The amount of hydrogen ions added by this change in P_{CO_2} is represented by arrow *2,* and the total base excess is the sum of the distances represented by arrows *1* and *2,* or 10 millimoles per liter.

If blood represented by point *B* were to be titrated with strong acid back to pH 7.40 at a P_{CO_2} of 40 mm Hg, 2 millimoles of acid per liter would be consumed in titrating buffers of the blood from pH 7.50 to pH 7.40, and 8 millimoles would be consumed in combining with the additional bicarbonate. This is also equal to the sum of the distances represented by the sum of arrows *1* and *2,* and therefore the amount of base excess can be

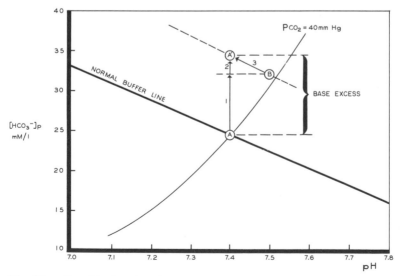

Fig. 23.   Graphical estimation of base excess.

estimated by dropping a vertical line from point B to the normal buffer line and measuring its length in millimoles per liter.

When acids such as ketone acids are added to blood, their hydrogen ions enter into two reactions: some combine with buffers of the blood, titrating them in the acid direction; and the rest combine with bicarbonate ions to form carbonic acid. The carbonic acid is dehydrated to carbon dioxide which is expired. Because some hydrogen ions added to blood combine with blood buffers, the number of hydrogen ions combining with bicarbonate ions is less than the total number added. Therefore, the decrease in bicarbonate concentration is less than the base deficit. For this reason, the amount of the base deficit is estimated on the pH-bicarbonate diagram by the vertical distance between the normal buffer line and the buffer line after acid has been added, not by the difference between the bicarbonate concentrations of normal and acidic blood.

> Example 17.   A sample of blood taken from a patient in diabetic acidosis is found to have a pH of 7.25 and a plasma bicarbonate concentration of 15 millimoles per liter. What is the base deficit?
>
> *First Step.*   The point is plotted as point B in figure 24. The normal point is assumed to be A. The vertical distance between A and B, represented by arrow *1*, is 9.0 millimoles per liter. This much bicarbonate has been displaced by added acid.

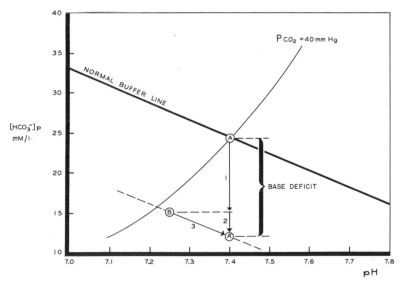

Fig. 24.   Graphical estimation of base deficit.

Second Step. The pH of the plasma is lower than normal, and this means that the buffers of the blood have been titrated in the acid direction. The amount of hydrogen ions which combined with the buffers of the blood can be estimated by titrating the blood in the alkaline direction to its normal pH. If this were done by removing carbonic acid, the blood would follow its buffer line to point *A'*, and the bicarbonate concentration would decrease. In order to change the blood from point *B* to *A'*, an amount of carbonic acid equivalent to arrow *2* would have to be removed. This is 3.0 millimoles per liter, and the total base deficit of the blood is the sum of arrows *1* and *2*, or 12.0 millimoles per liter.

The accuracy of these calculations of base excess or base deficit depends upon the assumption that the buffer line of the blood being studied is parallel to the normal buffer line. This assumption may not be correct, and to the extent that it is incorrect an error is introduced into the calculation. The following example shows the nature of the error and its possible magnitude.

Consider the blood represented by point *A* in figure 25. The pH is 7.47, and the plasma bicarbonate concentration is 38.0 millimoles per liter. We assume that its buffer line is parallel to

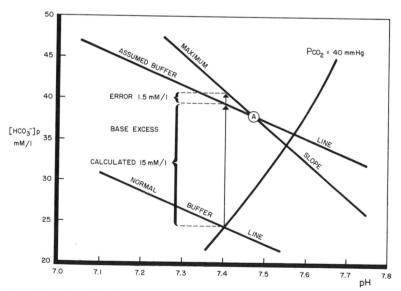

Fig. 25. Graphical calculation of base excess in blood showing the error if the true slope of the buffer line is that of blood containing 25 grams per cent of hemoglobin instead of being the slope of the normal buffer line.

the normal buffer line, and we draw a line with this slope through point *A*. The vertical distance between this line and the normal buffer line gives a calculated base excess of 15 millimoles per liter. However, the true slope of the buffer line of this particular sample of blood is greater than the assumed slope. Its true value is equal to that of blood containing 25 grams of hemoglobin per 100 ml which is very nearly the maximum slope likely to be encountered. This line, labeled "Maximum Slope," is also drawn through point *A*. To return blood represented by this line to a pH of 7.40 at a P_{CO_2} of 40 mm Hg would require 16.5 millimoles of strong acid, and therefore the previous estimation of base excess is in error by 1.5 millimoles per liter.

Point *B* in figure 26 represents blood whose pH is 7.30 and whose plasma bicarbonate concentration is 12 millimoles per liter. When the buffer line having a slope equal to that of normal blood is drawn through the point the calculated base deficit, measured as the vertical distance between this line and the normal line, is −15 millimoles per liter. Suppose, however, that the slope of the buffer line of this sample of blood is actually less than that of the normal line. The extreme is a slope equal to that of separated plasma or of blood containing zero hemoglobin, and a line having such a slope is drawn through point *B*. If this were

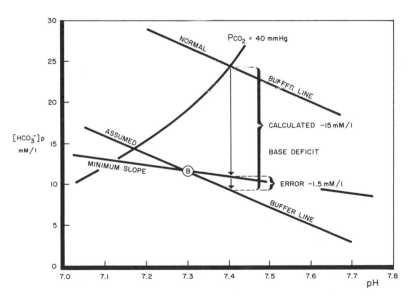

Fig. 26.   Graphical calculation of base deficit in blood *in vitro* showing the magnitude of the error if the true slope of the buffer line is that of separated plasma, or blood containing no hemoglobin, instead of being the slope of the normal *in vitro* buffer line.

the true buffer line of the blood, the error in estimating base deficit introduced by the use of the slope of the normal buffer line would be -1.5 millimoles per liter.

Base excess or base deficit is measured using blood *in vitro*. Section 2.1 explains that the actual slope of the buffer line of blood *in vivo* may be quite different from the slope of the buffer line of the blood of the same subject measured *in vitro*. Consequently, an error of unknown magnitude occurs if base excess or deficit accurately measured on blood *in vitro* is assumed to be the base excess or deficit of the same blood circulating within the subject.

The reader should remember two points:

1. Estimation of base excess or deficit is subject to error. However, the error may be within the limits of reliability of the data; and the error may also be within the limits set by variability of normal values (see section 2.5).
2. No sensible person would act upon a decision about a patient's base excess or deficit if correctness of the decision could be upset by errors in the way the numerical value for base excess or deficit was determined or calculated.

2.4. Normal Acid-Base Paths without Compensation

Changes in the acid-base pattern of the blood described in earlier sections can be produced in a normal man. The P_{CO_2} of the blood can be reduced by voluntary hyperventilation; this produces *respiratory alkalosis*. The P_{CO_2} of the blood can be raised by administration of a gas mixture high in carbon dioxide; this produces *respiratory acidosis*. *Metabolic alkalosis* can be produced by administration of sodium bicarbonate, and *metabolic acidosis* can be produced by giving ammonium chloride. Respiratory alkalosis or acidosis can be superimposed upon metabolic alkalosis or acidosis.

> Example 18.   A sample of arterial blood was taken from a normal man at rest. The man then hyperventilated by breathing as deeply as possible at the rate of 1 breath every 2 seconds for 2 minutes. At the end of the period of hyperventilation another sample of arterial blood was drawn. After the subject had recovered, he breathed a gas mixture of 7% carbon dioxide and 93% oxygen. At the end of 4 minutes another arterial blood sample was taken.
>
> The pH of each sample was measured. True plasma was separated from the blood and analyzed for total carbon dioxide. The bicarbonate concentration of the plasma and the P_{CO_2} of each sample was calculated. The data are given in table 9.

Table 9

Plasma Sample	At Rest (A)	Hyperventilating (B)	Breathing CO_2 (C)
pH	7.42	7.62	7.36
Total CO_2, mM/l	26.0	20.5	27.4
$[HCO_3^-]_p$, mM/l	24.8	19.9	26.0
P_{CO_2}, mm Hg	39.0	20.0	47.4

The results are plotted as points *A, B,* and *C* in figure 27. The arrow from point *A* to *B* represents the results of reducing the P_{CO_2} by hyperventilation. The blood of the subject moved down its normal *in vivo* buffer line, and at point *B* the subject was in *uncompensated respiratory alkalosis.* The characteristics of this condition are low P_{CO_2}, high pH, and low plasma bicarbonate concentration.

The arrow from *A* to *C* represents the results of increasing the P_{CO_2}. The blood of the subject moved up its normal *in vivo* buffer line, and at *C* the subject was in *uncompensated respiratory acidosis.* The characteristics of this condition are high P_{CO_2}, low pH, and high plasma bicarbonate concentration.

Example 19. The same subject took 30 grams of sodium bicarbonate by mouth. Two hours later a sample of arterial blood was taken. The procedures of hyperventilation and

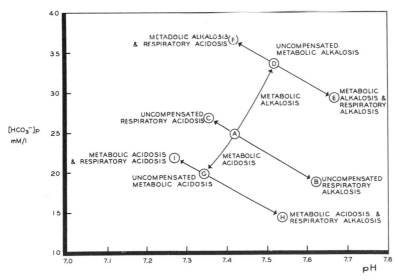

Fig. 27.   Acid-base paths *in vivo.*

breathing carbon dioxide were repeated, and blood samples were taken. The analytical data are given in table 10.

Table 10

Plasma Sample	At Rest (D)	Hyperventilating (E)	Breathing CO_2 (F)
pH	7.52	7.67	7.42
Total CO_2, mM/l	35.0	30.4	38.4
$[HCO_3^-]_p$, mM/l	33.7	29.6	36.5
P_{CO_2}, mm Hg	42.0	26.4	57.0

The points are plotted as D, E, and F in figure 27. The arrow from point A to D represents the development of metabolic alkalosis, and D represents the state of *uncompensated metabolic alkalosis*. The characteristics of this condition are normal P_{CO_2}, high pH, and high plasma bicarbonate concentration.

In going from A to D the blood of the subject moved from its normal *in vivo* buffer line to one higher than normal but parallel to it. This is shown by the effects of hyperventilation and breathing carbon dioxide. At E the subject had superimposed a respiratory alkalosis upon his initial metabolic alkalosis. On account of increased elimination of carbon dioxide, the P_{CO_2} fell, and the blood moved down the *in vivo* buffer line to a point characterized by higher pH and lower bicarbonate. In this condition the total carbon dioxide of the blood approached normal value; but, because the blood was moving down the *in vivo* buffer line, the pH increased still further.

On the other hand, decreased elimination of carbon dioxide superimposed a respiratory acidosis upon the initial metabolic alkalosis. The P_{CO_2} rose, and the blood moved up the *in vivo* buffer line to point F characterized by lower pH and still higher bicarbonate. In this condition the pH of the blood was normal despite the fact that the total carbon dioxide of the blood was high.

Example 20.   The same subject, after returning to normal, took 15 grams of ammonium chloride by mouth. Two hours later a resting sample of arterial blood was taken. The procedures of hyperventilation and breathing carbon dioxide were repeated, and blood samples were obtained. True plasma was analyzed for pH and total carbon dioxide, and plasma bicarbonate concentration and P_{CO_2} were calculated. The data are given in table 11.

Table 11

Plasma Sample	At Rest (G)	Hyperventilating (H)	Breathing CO_2 (I)
pH	7.35	7.54	7.27
Total CO_2, mM/l	20.9	14.9	24.0
$[HCO_3^-]_p$, mM/l	19.8	14.4	22.5
P_{CO_2}, mm Hg	37.0	17.4	50.3

   The points are plotted as G, H, and I in figure 27. The arrow from A to G represents the development of metabolic acidosis. At point G the subject was in *uncompensated metabolic acidosis*. The characteristics of this condition are normal P_{CO_2}, low pH, and low plasma bicarbonate concentration.

   In going from A to G the blood of the subject moved from its normal *in vivo* buffer line to a line lower than and parallel to the normal one. This is shown by the effects of hyperventilation and breathing carbon dioxide. At point I the subject had added a respiratory acidosis to his initial metabolic acidosis. Because elimination of carbon dioxide was reduced, the P_{CO_2} of the blood was increased, and the blood moved up its *in vivo* buffer line to a point characterized by lower pH and higher bicarbonate concentration. Despite the fact that the bicarbonate concentration approached normal, the pH of the blood was very low.

   On the other hand, hyperventilation reduced the P_{CO_2} of the blood and added respiratory alkalosis to the initial metabolic acidosis. The blood moved down the *in vivo* buffer line to a point at which the pH was higher than normal, but the bicarbonate concentration was far below normal.

   The points on figure 27 illustrate the nature of the areas on a pH-bicarbonate diagram:

1. Any condition represented by a point falling within the area above and to the left of the P_{CO_2} equals 40 mm Hg isobar has a component of *respiratory acidosis*.
2. Any condition represented by a point falling within the area below and to the right of the P_{CO_2} equals 40 mm Hg isobar has a component of *respiratory alkalosis*.
3. Any condition represented by a point falling to the left of the normal pH has a *low* pH.
4. Any condition represented by a point falling to the right of the normal pH has a *high* pH.
5. Any condition represented by a point falling within the area above the normal buffer line has a component of *base excess*.

6. Any condition represented by a point falling within the area below the normal buffer line has a component of *base deficit*.

2.5. Normal Ranges

To make intelligent use of acid-base data, one must know the limits within which normal values are likely to fall.

Several sets of observations on persons living at sea level and apparently forming a representative sample of the normal population show that 95 per cent or more of the values of arterial or arterialized blood are within the following limits:

$$pH - 7.35-7.45$$
$$[HCO_3^-]_p \text{ (mM per liter)} - 23-28 \text{ (women lower than men)}$$
$$P_{CO_2} \text{ (mm Hg)} - 35-48$$

On the average, the plasma bicarbonate concentrations of women are about one millimole per liter lower than those of men. There are some persons apparently free of disease who are outside the normal limits. Some persons show no day-to-day variations, whereas others may have differences as great as 4 millimoles per liter between plasma bicarbonate concentrations measured on different days. There are no obvious reasons for such variations.

The data quoted are for persons living at sea level. At higher altitudes the reduced partial pressure of oxygen in the inspired air reduces the P_{O_2} of alveolar air and arterial blood.

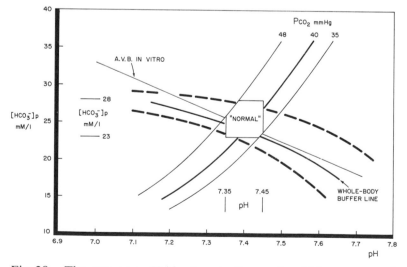

Fig. 28.   The area on a pH-bicarbonate diagram in which most normal values fall and the areas of normal response to acute hypercapnia and hypocapnia.

This stimulates respiration, and the consequent hyperventilation reduces the P_{CO_2} of arterial blood. At Salt Lake City, for example, where much of the population lives at 4,500 feet and where the barometric pressure is about 650 mm Hg the normal resting arterial P_{CO_2} is from 32 to 36 mm Hg. Renal compensation for this mild chronic respiratory alkalosis is complete, and arterial pH is within normal limits. Plasma bicarbonate is therefore well below that found at sea level, the average value being about 20 millimoles per liter. Populations at other altitudes have corresponding deviations from "normal."

The data from which acid-base values are calculated are obtained by chemical and physical means which are subject to error. All surveys of clinical laboratories have shown that even the best occasionally make mistakes and that the worst frequently report wildly erroneous results. Consequently, anyone using and interpreting acid-base data must know how the data are obtained and to what extent he can depend upon them.

2.6. Chemical Regulation of Respiration

In a normal man at rest breathing atmospheric air at a barometric pressure of 760 mm Hg, respiration is regulated to maintain a constant rate of ventilation of the alveolar spaces. At the normal rate of ventilation, alveolar P_{CO_2} is 40 mm Hg; and since arterial blood is in equilibrium with alveolar carbon dioxide, the P_{CO_2} of arterial blood is also 40 mm Hg. At the same rate of ventilation, alveolar P_{O_2} is about 100 mm Hg, and the P_{O_2} of arterial blood is a few mm Hg lower. Arterial blood containing carbon dioxide and oxygen at these partial pressures bathes the chemoreceptors in the aortic and carotid bodies, and it perfuses the brain.

The P_{CO_2} of arterial blood is the most important chemical factor governing respiration. An increase in alveolar P_{CO_2}, and hence an increase in arterial P_{CO_2}, stimulates respiration so that the rate of ventilation of the alveolar spaces is increased, and a decrease in alveolar and arterial P_{CO_2} inhibits respiration so that ventilation is decreased.

Carbon dioxide and hydrogen ions are related through the reactions

$$CO_2 + H_2O \; \rightleftharpoons \; H_2CO_3 \; \rightleftharpoons \; H^+ + HCO_3^-. \qquad (70)$$

If the P_{CO_2} rises, $[H^+]$ increases (or pH falls), and if the P_{CO_2} falls, $[H^+]$ decreases (or pH rises). The magnitude of the change in $[H^+]$ following a given change in P_{CO_2} is, of course, determined by the buffer value of the medium in which the change occurs.

In man an acute increase in P_{CO_2} of 1 mm Hg causes a rise in $[H^+]_p$ of 0.74 to 0.77 nanomoles per liter, and in a chronically hypercapnic man the increase in $[H^+]_p$ for a similar 1 mm Hg rise in P_{CO_2} is 0.24 nanomoles per liter. Cerebrospinal fluid is more poorly buffered than blood, and in it the value of $\Delta[H^+]/\Delta P_{CO_2}$ lies between 0.9 and 1.0 nanomoles per liter.

Because carbon dioxide is highly soluble in water and in the lipids which make up much of cell membranes, any change in P_{CO_2} of arterial blood supplying a part is rapidly followed by a corresponding change in the P_{CO_2} of the part's interstitial fluid and of the interior of its cells. Consequently, $[H^+]$ of interstitial and intracellular fluids is quickly altered by changes in alveolar P_{CO_2}.

About half the effect of changes in P_{CO_2} upon respiration is exerted through the effect of carbon dioxide upon $[H^+]$ of interstitial fluid of the brain. Respiratory chemoreceptors lie near the ventral surface of the medulla. They are bathed by interstitial fluid which is derived from cerebrospinal fluid. These chemoreceptors appear to be sensitive to $[H^+]$ of their environment. If $[H^+]$ rises, they respond by stimulating the respiratory center, causing an increase in alveolar ventilation. If $[H^+]$ falls, chemoreceptor drive of the respiratory center is reduced, and alveolar ventilation falls. Although the neural tissue containing the chemoreceptors is perfused by blood, there is a barrier between the blood in the capillaries and the interstitial fluid of the brain. The barrier prevents the passage of the buffers of the blood from blood to cerebral interstitial fluid. Because carbon dioxide, on account of its lipid solubility, can readily cross the barrier, and because cerebral interstitial fluid is poorly buffered, a change in arterial P_{CO_2} is rapidly followed by a relatively large change in $[H^+]$ of the fluid surrounding the respiratory chemoreceptors. Therefore, an increase in alveolar and arterial P_{CO_2} increases cerebral interstitial $[H^+]$; chemoreceptors respond by driving the respiratory center; and alveolar ventilation increases. A fall in alveolar and arterial P_{CO_2} decreases cerebral interstitial $[H^+]$; chemoreceptor drive of the respiratory center is reduced; and alveolar ventilation is diminished.

On account of the barrier between cerebral capillaries and interstitial fluid, there is no immediate exchange of bicarbonate between blood and cerebral interstitial fluid. The bicarbonate concentration of cerebral interstitial fluid is slowly adjusted by the tissues which manufacture the cerebrospinal fluid. If the P_{CO_2} and the $[H^+]$ of cerebral tissue increase, $[HCO_3^-]$ of cerebral spinal and cerebral interstitial fluid slowly rises. The means by which this is accomplished are unknown, but the process may be similar to that by which the kidney responds to changes in P_{CO_2}. A rise in P_{CO_2} causes renal tubular cells to increase their rate of secretion of hydrogen ions into the tubular urine and the rate at which they return bicarbonate to the blood. In the brain an increase in P_{CO_2} might increase the rate at which the choroid plexus secretes hydrogen ions into the blood and returns bicarbonate ions to the cerebrospinal fluid.

As the result of the slow removal of hydrogen ions and addition of bicarbonate ions, $[H^+]$ of cerebrospinal and interstitial fluids decreases over a period of hours or days although the P_{CO_2} remains elevated. Since the chemoreceptors respond to $[H^+]$ and not to P_{CO_2}, the fall in $[H^+]$ removes some of the stimulus for ventilation, and respiratory minute volume returns toward normal. On the other hand, if the P_{CO_2} is below normal, the tissues forming cerebrospinal fluid allow hydrogen ions to accumulate within the fluid. Consequently, $[H^+]$ of cerebrospinal and interstitial fluids slowly rises, and $[HCO_3^-]$ of the fluids slowly falls. Although the P_{CO_2} remains low, the return of cerebral interstitial $[H^+]$ toward normal removes to some extent the inhibition of respiration produced by the initial reduction of P_{CO_2}. Alveolar ventilation then rises toward normal.

The other half of the effect of P_{CO_2} upon respiration is exerted through action of carbon dioxide upon peripheral chemoreceptors, those in the aortic and carotid bodies and perhaps elsewhere. The effect is the same as that of P_{CO_2} upon the central receptor system: an increase in P_{CO_2} stimulates respiration and increases alveolar ventilation; a decrease in P_{CO_2} inhibits respiration and reduces alveolar ventilation.

The partial pressure of carbon dioxide has important additional effects. When arterial P_{CO_2} rises acutely above 70 to 80 mm Hg, carbon dioxide depresses the central nervous system. As the P_{CO_2} rises, the subject becomes somnolent and confused, and he may lapse into unconsciousness. Pressures above 200 mm Hg are anesthetic and convulsant. Acute reduction of P_{CO_2}, by increasing pH and decreasing the ionization of calcium, causes increased neuromuscular excitability which may result in tetany.

The hydrogen ion concentration of arterial blood can be altered by addition of non-volatile acid or of base as well as by changes in P_{CO_2}. Such changes in $[H^+]$ in themselves affect respiration, partly through the effect of $[H^+]$ upon peripheral chemoreceptors and partly through the effect of $[H^+]$ upon central chemoreceptors whose locus and nature are poorly understood. The effect upon ventilation is the same as that produced by alteration in $[H^+]$ caused by changes in P_{CO_2}: an increase in $[H^+]$ at constant P_{CO_2} stimulates respiration so that

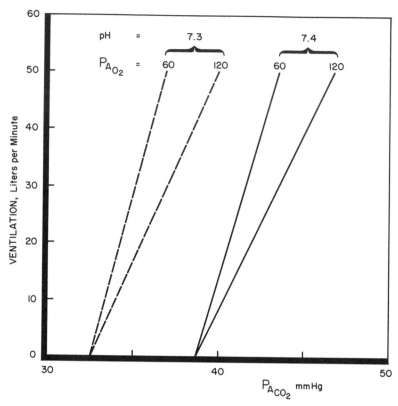

Fig. 29.   The relation between ventilation of the lungs and the alveolar
P_{CO_2} in human subjects. The two lines on the right show the relation
obtaining when the pH of the plasma has the normal value of 7.4 at
rest and when the alveolar P_{O_2} is at the hypoxic level of 60 mm Hg or
at the supernormal value of 120 mm Hg. The two lines at the left show
the same relation in a subject whose plasma pH at rest is 7.3 as the
result of prolonged ingestion of ammonium chloride. Adapted from
Cunningham, Shaw, Lahiri and Lloyd, 1961, *Quart. J. Exp. Physiol.*
46:323. (Reproduced by permission.)

alveolar ventilation increases, and a decrease in $[H^+]$ at constant
P_{CO_2} inhibits respiration so that ventilation decreases. When both
$[H^+]$ and P_{CO_2} change, ventilation responds as though the effects
of the two were algebraically added.

The chemoreceptors of the carotid and aortic bodies respond to a reduction in the partial pressure of oxygen in arterial
blood by increasing the frequency with which they send impulses
over afferent nerves to the respiratory center. Increasing frequency stimulates respiration so that alveolar ventilation increases. When the P_{O_2} of arterial blood is in the normal range
of 90 to 100 mm Hg, the chemoreceptors discharge at a low
frequency. A sudden increase in arterial P_{O_2} to about 200 mm Hg

stops chemoreceptor discharge, and alveolar ventilation falls by about 10 per cent. This fall in turn causes retention of carbon dioxide. The consequent slight rise in arterial P_{CO_2} stimulates respiration, and ventilation of the lungs returns very nearly to the level obtaining before P_{O_2} was increased. On the other hand, a decrease in arterial P_{O_2} below normal strongly stimulates the carotid and aortic chemoreceptors, and the frequency of impulses reaching the respiratory center from the chemoreceptors is greatly increased. Respiration is stimulated. Increased alveolar ventilation reduces arterial P_{CO_2}, and this inhibits respiration. Consequently, there is a balance between the reflex drive of respiration initiated by the chemoreceptors' response to low P_{O_2} and the inhibition resulting from respiratory alkalosis. The two opposing influences cancel each other so that there is very little increase in alveolar ventilation until arterial P_{O_2} falls below 60 mm Hg at which P_{O_2} hemoglobin is 85 to 90 per cent saturated. Below this P_{O_2} reflex drive of ventilation by chemoreceptor response to hypoxia dominates, and ventilation is increased. Although hypoxia itself depresses the central nervous system including the respiratory center, strong reflex drive of respiration originating in the peripheral chemoreceptors can maintain ventilation of the lungs when the P_{O_2} of arterial blood is chronically low.

The effects of oxygen and carbon dioxide are complexly interrelated. Reduction of P_{O_2} below the normal value increases the sensitivity of the respiratory mechanism to changes in P_{CO_2}. When arterial P_{O_2} is 60 mm Hg, the ventilatory response to a given increase in P_{CO_2} is about double that occurring when the P_{O_2} is 120 mm Hg. On the other hand, an increase in arterial P_{CO_2} above the normal level of 40 mm Hg increases the ventilatory response to hypoxia.

2.7. Respiratory Compensation for Metabolic Alkalosis or Acidosis

Respiratory compensation for metabolic alkalosis or acidosis is the result of alterations in ventilation of the lungs caused by changes in the pH of arterial blood.

Suppose that metabolic alkalosis develops, and the pH of arterial blood increases. Increased pH depresses respiration and reduces ventilation of alveolar spaces, and the P_{CO_2} of alveolar air and arterial blood rises. The increase in P_{CO_2} titrates the blood along its buffer line in the direction of lower pH and higher bicarbonate concentration. Reduction of pH by increased P_{CO_2} is respiratory compensation for metabolic alkalosis.

The sequence of events is shown in figure 30. Note that in order to give greater detail to the graph, the units of the coordinates are smaller than those of previous diagrams. Development of metabolic alkalosis is represented by the arrow from the normal point A to B. For the sake of simplicity of argument, it is assumed that, as metabolic alkalosis is developing, respiration remains unchanged so that point B lies on the P_{CO_2} equals 40 mm Hg isobar. A change from A to B increases the pH, and the increase in pH depresses respiration. Therefore, retention of carbon dioxide moves the blood from B to C. This process stops when depression of respiration by increasing pH is exactly balanced by stimulation of respiration resulting from increased P_{CO_2}. Of course, both processes—development of metabolic alkalosis and its respiratory compensation—occur together, and the actual path is the resultant of the two changes as shown by the broken line from A to C.

Respiratory compensation for metabolic alkalosis cannot be so complete as to bring the pH to its normal value. As the pH returns toward normal, depression of respiration caused by increased pH disappears, and if the pH were to return to normal, respiratory depression would be zero. This leaves respiratory stimulation by increased P_{CO_2} unopposed. The consequent stimulation of respiration would lower the P_{CO_2} and drive the blood down its buffer line in the direction of increased pH until depres-

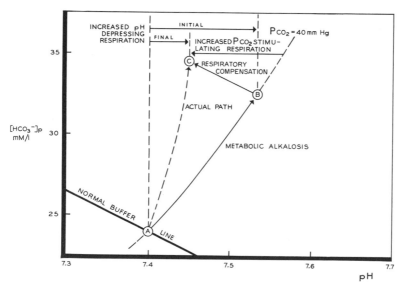

Fig. 30.   Processes occurring in respiratory compensation for metabolic alkalosis.

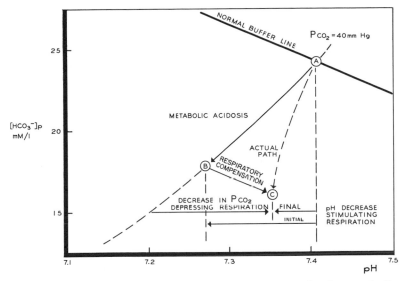

Fig. 31.   Processes occurring in respiratory compensation for metabolic acidosis.

sion caused by increasing pH exactly balances the decreasing stimulus of high P_{CO_2}.

Respiratory compensation for metabolic acidosis occurs because decreased pH produced by metabolic acidosis stimulates respiration and increases alveolar ventilation. Alveolar P_{CO_2} and arterial P_{CO_2} are reduced, and this titrates the blood along its buffer line in the direction of higher pH and lower bicarbonate concentration.

The sequence of events is shown in figure 31, where again it is assumed that metabolic acidosis does not affect respiration until point B on the P_{CO_2} equals 40 mm Hg isobar is reached. The change from A to B decreases the pH, and decreased pH stimulates respiration. Respiratory compensation moves the blood from B to C. The process stops when stimulation of respiration by low pH is balanced by depression caused by decreased P_{CO_2}, and compensation is not complete. Since development of metabolic acidosis and its compensation occur together, the actual path is the resultant of the two changes shown by the broken arrow from A to C.

2.8. Renal Processes Responding to Acid-Base Changes

The function of the kidney is to maintain the constancy of the internal environment. By regulating the rate of water loss in the urine, it defends the internal environment against excessive hydration or dehydration and keeps the osmotic pressure of the blood at a value equal to that of a solution containing 154 millimoles of NaCl per liter. By regulating excretion of individual ions, it maintains the normal electrolyte pattern of plasma and interstitial fluid. By regulating the acidity of the urine and the rate of excretion of electrolytes, it helps to keep the pH of plasma within normal limits. The separate functions of the kidney are integrated, and, in the face of pathological disturbances, one function may be sacrificed for another function.

Urine can be either acid or alkaline. When the urine is acid, the acid excreted is removed from the blood, and in effect an equal quantity of base is added to the blood.

Renal tubules secrete hydrogen ions into the tubular urine. The reactions by which this is accomplished are unknown, but, whatever the process, one hydroxyl ion (OH^-) is left within the renal tubular cells for each hydrogen ion secreted. The accumulation of hydroxyl ions within tubular cells raises their internal pH, and the reactions

$$CO_2 + H_2O \rightarrow H_2CO_3 \rightarrow HCO_3^- + H^+ \qquad (71)$$

go to the right within the cells. Hydrogen ions formed from carbon dioxide and water combine with hydroxyl ions. This process of intracellular neutralization by means of carbon dioxide keeps the pH of the renal tubular cells within the limit necessary for the secretory process to continue. Bicarbonate ions, which are the by-product of intracellular neutralization, are transported into the peritubular fluid and renal venous plasma with the result that, for every mole of acid secreted into tubular urine, one new mole of bicarbonate (or base) appears in blood.

Hydration of carbon dioxide within the kidney is catalyzed by carbonic anhydrase. If renal carbonic anhydrase is inhibited as, for example, by acetazolamide, renal tubular secretion of acid may be reduced or abolished. However, hydration of carbon dioxide can also occur without enzymatic catalysis, and under some conditions, particularly severe metabolic acidosis, intracellular neutralization and renal acid secretion can occur without the assistance of carbonic anhydrase.

When acid is secreted by the tubular cells, a balance of electrical charges must be maintained. Apparently this is accomplished by the forced exchange of sodium ions for hydrogen ions so that, for every hydrogen ion secreted into the tubular urine, one sodium ion derived from the glomerular filtrate is reabsorbed and transported into the peritubular fluid and renal venous plasma. Since a bicarbonate ion is simultaneously transported in the same direction, secretion of acid into the tubular urine is accompanied by the addition of sodium bicarbonate to the venous blood. The sodium ion transported into venous blood was originally derived by glomerular filtration of arterial blood, so this process conserves sodium.

Renal tubular cells also secrete potassium ions into the tubular urine, and secretion of potassium ions and secretion of hydrogen ions are interrelated. An increased rate of secretion of one ion is accompanied by a decreased rate of secretion of the other, and if for any reason secretion of one is diminished, secretion of the other is usually enhanced. In respiratory acidosis, the rate of acid secretion is high, and the rate of potassium secretion is depressed. Consequently, potassium is conserved, and the body's load of potassium tends to increase. During potassium deficiency, on the other hand, renal secretion of potassium is reduced; secretion of hydrogen ions into the urine increases with the result that metabolic alkalosis of renal origin develops.

The minimum pH of the urine is about 4.5. This is a hydrogen ion concentration of 0.000,03 mole per liter, and it is approximately 800 times more acid than plasma. The secretory process is not capable of establishing a steeper gradient. As a consequence, the rate of excretion of acid is proportional to the rate of excretion of buffer in the urine. When the buffer capacity of tubular urine is low, secretion of only a small amount of acid into tubular urine reduces its pH to 4.5. When the buffer capacity is high, a large amount of acid can be secreted into the urine before its pH reaches the limiting value of 4.5.

Under normal circumstances, phosphate is one of the two important buffers in the urine. The glomerular filtrate contains phosphate and sodium ions in concentrations very nearly equal to their concentrations in plasma. The pH of the glomerular filtrate is the same as that of plasma, about pH 7.4. The second hydrogen of phosphoric acid dissociates according to the equation

$$H_2PO_4^- \rightleftharpoons HPO_4^= + H^+. \tag{72}$$

This dissociation can be expressed in the logarithmic form of the mass-action equation, which can be derived by the methods explained in section 1.13:

$$pH = pK + \log \frac{[HPO_4^=]}{[H_2PO_4^-]} \tag{73}$$

The pK is 6.8, and at the normal pH of plasma the equation becomes

$$7.4 = 6.8 + \log \frac{[HPO_4^=]}{[H_2PO_4^-]}. \tag{74}$$

The equation can be solved for the ratio of the dibasic to monobasic phosphate. Transposing, we have

$$\log \frac{[HPO_4^=]}{[H_2PO_4^-]} = 0.6, \tag{75}$$

$$\frac{[HPO_4^=]}{[H_2PO_4^-]} = \text{antilog } 0.6. \tag{76}$$

The antilogarithm of 0.6 is 4, and the ratio is 4:1. This means that there is four times as much dibasic as monobasic phosphate in the glomerular filtrate. Out of every five phosphate molecules appearing in the glomerular filtrate, four are dibasic and have two negative charges, and one is monobasic and has one negative charge. Five phosphate molecules have a total of nine negative charges which must be balanced by positive charges, chiefly supplied by sodium. If the five phosphate molecules were to be excreted unchanged, a total of nine sodium ions would also have to be excreted.

When hydrogen ions are added to tubular urine, the dissociation of phosphate as expressed in equation (72) is driven to the left, and monobasic phosphate is formed from dibasic phosphate. If enough acid is added to tubular urine to reduce its pH to 4.5, almost all the phosphate is converted to the monobasic form. Substitution of 4.5 in equation (73) gives a ratio of dibasic to monobasic phosphate of 1:200. This means that more than 99 per cent of the phosphate molecules in the urine have only one charge apiece. Only five sodium ions are required to balance their negative charges, and, for every five phosphate molecules appearing in acid urine, four sodium ions can be saved and returned to the blood. These processes are summarized in figure 32.

The second buffer which allows conservation of sodium is ammonia. Ammonia (NH_3) is formed within renal tubular cells from glutamine and some amino acids, and ammonia (NH_3) but not ammonium (NH_4^+) diffuses from the cells into the tubular urine. There it reacts with hydrogen ions to give ammonium (NH_4^+) ions. Because ammonia has picked up hydrogen ions from the urine, those hydrogen ions do not contribute to the acidity of the urine. Consequently, as long as ammonia (NH_3) is added to tubular urine, hydrogen ion secretion and its accompanying sodium reabsorption can continue. These processes are summarized in figure 33.

The net amount of acid eliminated by the kidney in any period of time is obtained by titrating the urine excreted from its acid pH back to the pH of the blood from which the urine was derived and by adding to this quantity the amount of ammonium ions found in the urine. The first part of this sum gives the number of hydrogen ions buffered by phosphate and similar buffers,

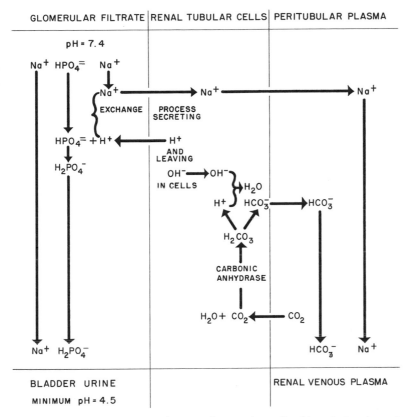

Fig. 32.   Scheme representing renal secretion of acid and titration of phosphate.

and it is the *titratable acidity* of the urine. The second part gives the number of hydrogen ions buffered by ammonia. The sum is the *total acid excretion,* and it measures the amount of base added to the blood by the kidney in the time the urine was formed.

Renal acid secretion has two functions: to add base to the blood as described above and to prevent bicarbonate contained in the glomerular filtrate from being excreted. The bicarbonate concentration of the glomerular filtrate is nearly the same as that of plasma, and with each bicarbonate anion is associated a sodium cation. When sodium and bicarbonate of the glomerular filtrate reach a segment of the tubules where acid is secreted, the sodium is reabsorbed, and hydrogen ions combine with the bicarbonate to form carbonic acid. The carbonic acid is dehydrated to give carbon dioxide and water, and the P_{CO_2} of the tubular urine increases. Most of the carbon dioxide diffuses through the tubular cells to the blood, and the rest goes into the bladder urine. By

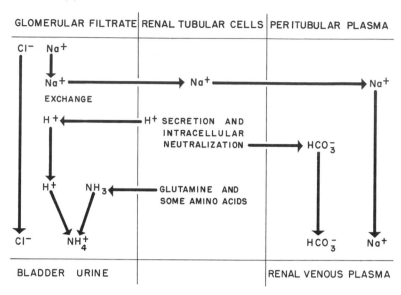

GLOMERULAR FILTRATE | RENAL TUBULAR CELLS | PERITUBULAR PLASMA

BLADDER   URINE | | RENAL VENOUS PLASMA

Fig. 33.   Scheme representing renal excretion of ammonium ions.

this series of reactions, bicarbonate disappears from the urine, and an equal amount, formed within the tubular cells during acid secretion, appears in the blood. The total process (fig. 34) is equivalent to reabsorption of bicarbonate, although the fact that the bicarbonate molecules appearing in renal venous blood are not actually the same molecules as were filtered in the glomeruli makes the term "bicarbonate reabsorption" inaccurate.

Because acid is used in bicarbonate reabsorption, there is an inverse relation between the rate of bicarbonate filtration and the rate of excretion of titratable acid. In normal man the rate of glomerular filtration is approximately constant. Consequently, the rate at which bicarbonate is filtered and presented to the tubules is chiefly a function of its plasma concentration. If the plasma concentration is normal, 24 millimoles per liter, and if the rate of glomerular filtration is 0.125 liters per minute, then 3 millimoles of bicarbonate are filtered per minute. If the plasma bicarbonate is low, say 12 millimoles per liter, then at the same rate of glomerular filtration only 1.5 millimoles are filtered per minute. If the plasma bicarbonate is high at 36 millimoles per liter, 4.5 millimoles of bicarbonate are filtered per minute. In a normal man having a P_{CO_2} of 40 mm Hg the rate of acid secretion by his renal tubules is approximately constant at about 3.5 millimoles per minute. All the bicarbonate can be reabsorbed, leaving 0.5 millimole of acid to be excreted in the urine. When

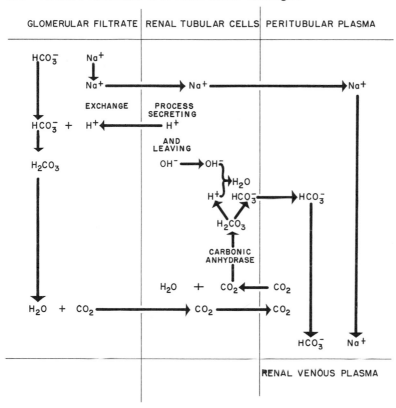

Fig. 34.   Scheme representing renal reabsorption of bicarbonate by means of acid secretion.

the plasma bicarbonate is 12 millimoles per liter, 3.5 minus 1.5 or 2 millimoles of acid are left over after all the bicarbonate has been reabsorbed. However, when the plasma bicarbonate is above 28 millimoles per liter, bicarbonate is filtered at a greater rate than acid is secreted, and, after all the acid has been used up in reabsorbing bicarbonate, some bicarbonate remains in the tubular urine to pass into the bladder. With plasma bicarbonate of 36 millimoles per liter, 4.5 minus 3.5 or 1 millimole per minute would appear in bladder urine if the rate of glomerular filtration were 0.125 liter per minute and the rate of acid secretion were 3.5 millimoles per minute.

There is a reciprocal relation between renal excretion of bicarbonate and renal excretion of chloride. In general, when excretion of bicarbonate in urine is low, the major anion of the urine is chloride. When bicarbonate excretion is high, chloride excretion is reduced.

2.9. Renal Responses to Metabolic
Alkalosis and Acidosis

The two major variables affecting renal responses to metabolic alkalosis and acidosis are the plasma bicarbonate concentration and the amount of buffer in the urine.

In metabolic alkalosis the plasma pH is up; the plasma bicarbonate concentration is above normal; and if respiratory compensation has occurred the P_{CO_2} is raised. When the plasma bicarbonate concentration is greater than 28 millimoles per liter, the rate of glomerular filtration of bicarbonate is greater than the rate at which renal tubules can reabsorb bicarbonate. Consequently, bicarbonate passes into the urine, and an alkaline urine is excreted. To maintain electrical neutrality, cations, chiefly sodium, appear in the urine with bicarbonate ions, and plasma sodium concentration tends to decrease. Formation and excretion of alkaline urine, in effect, add acid to the blood, with the result that plasma pH and bicarbonate return toward normal values. On account of the reciprocal relation between bicarbonate excretion and that of chloride, the rate of chloride excretion is lowered, and plasma chloride concentration tends to rise. This increase in plasma chloride replaces bicarbonate which is lost. Finally, no ammonium ions are excreted in alkaline urine.

In metabolic acidosis, the plasma bicarbonate concentration and pH are both below normal. If respiratory compensation has occurred, the P_{CO_2} is also down. Because the plasma bicarbonate concentration is low, the rate at which bicarbonate is filtered by the glomeruli and presented to the renal tubules is much lower than the rate at which renal tubular cells secrete acid into tubular urine. All filtered bicarbonate is reabsorbed, and extra acid appears in the urine as titratable acidity. The lower the plasma bicarbonate, the more acid is left over to appear in the urine. Excretion of acid, in effect, adds base to the blood, and as a result of this renal response, both plasma bicarbonate and pH rise.

When the urine is acid, ammonium ions (NH_4^+) are excreted, and sodium ions are conserved by the processes outlined in figure 33. The rate of ammonium excretion depends on two factors: (1) the pH of the urine and (2) the duration of acidosis. The lower the urine pH, the more ammonium is excreted. In this relation the urine pH is the independent variable, and ammonium excretion is the dependent one. If the pH of the urine is lowered, the rate of ammonium excretion immediately increases, and if the pH is raised, ammonium excretion falls. The magnitude of the response depends on the second factor, duration of acidosis. The same qualitative relation between urine pH and ammonium excretion persists, but after a few days of severe acidosis more ammonium is excreted at a given pH. If, for example, a normal man excretes 30 millimoles of ammonium a day when his urine pH is 5, he may excrete 200 millimoles a day at the same pH if he has had severe metabolic acidosis for two or three days. However, if something is done to raise his urine pH abruptly to 7.5, his ammonium excretion drops to zero, just as it would if he were normal.

The relation between the amount of buffer in the urine and the excretion of titratable acidity is illustrated by diabetes mellitus. In this disease, fat metabolism increases; acetoacetic acid is formed in the liver and transported to other tissues to be oxidized. In uncontrolled diabetes mellitus acetoacetic acid may be formed at a rate much greater than it can be oxidized, and the extra amount is excreted in the urine. This may be as much as 500 to 1,000 millimoles per day. Formation and excretion of this load of acetoacetic acid imposes two problems on the acid-base regulating mechanisms: (1) the addition of acetoacetic acid to the blood causes metabolic acidosis, and (2) conservation of plasma cations, sodium and potassium, is difficult during renal excretion of the extra acid.

Acetoacetic acid is made from fat in the acid form shown on the left:

$$CH_3-\overset{\displaystyle O}{\overset{\|}{C}}-CH_2-COOH \tag{77}$$

$$\rightleftharpoons CH_3-\overset{\displaystyle O}{\overset{\|}{C}}-CH_2-COO^- + H^+.$$

The pK of acetoacetic acid is approximately 4, and at the pH of the blood it is completely ionized. Therefore, one hydrogen ion must be accepted by blood buffers for every molecule of acetoacetic acid in the blood, and this is the primary cause of metabolic acidosis in diabetes mellitus. The pH of the blood falls, and respiratory compensation begins. After the hydrogen ions are accepted by the blood buffers, the anions of acetoacetic acid are balanced electrically by sodium and potassium ions.

Acetoacetate in the plasma is filtered in the renal glomeruli. Some of it is reabsorbed by the renal tubules, and the rest passes into bladder urine. Because acetoacetate is in the anionic form ($-COO^-$), it is accompanied by cations. If all that is excreted into the urine remained in the anionic form, an equivalent amount of cations would be excreted, and with as much as 500 millimoles of acetoacetate excreted per day, the body's cation content would be rapidly and disasterously depleted. Acid secretion by renal tubules is a partial defense against depletion. An amount of base equal to the acid secreted is returned to the blood, and metabolic acidosis is corrected to the extent that acid is excreted. Aceto-acetate acts as a buffer in tubular urine, and, because buffer present there takes up hydrogen ions, excretion of titratable acid increases along with the rate of excretion of the buffer. Unfortunately, the pK of acetoacetic acid is about 4. This means that only one-fourth to one-third of the acetoacetate present in the urine can be titrated by secreted acid from the anionic form ($-COO^-$) to the acid form ($-COOH$) before the limiting urine pH of 4.5 is reached. The rest of the acetoacetate is excreted as anions, and it takes cations with it. This is the reason that acidosis, dehydration, and salt depletion continue in uncontrolled diabetes mellitus despite maximal operation of compensating and correcting processes.

2.10. Renal Compensation for Respiratory Alkalosis or Acidosis

Renal compensation for respiratory alkalosis or acidosis is the result of the effect of the P_{CO_2} of the blood upon the kidney. The experimentally determined fact is that the rate of reabsorption of bicarbonate by renal tubules is a direct and linear function of the P_{CO_2}. The relation between the two variables is shown in figure 35. The rate of reabsorption of bicarbonate in a normal man having a rate of glomerular filtration of 0.125 liters per minute is plotted as the dependent variable against plasma P_{CO_2}. As the P_{CO_2} rises, so does the rate of reabsorption, and as the P_{CO_2} falls, bicarbonate reabsorption falls with it. This dependence of bicarbonate reabsorption upon P_{CO_2} is a function of the P_{CO_2} itself, not of the pH, and it is a renal process operating in addition to the effects of urine buffer concentration, ammonia formation, and plasma bicarbonate concentration already described.

Renal tubular reabsorption of bicarbonate is chiefly the result of acid secretion by the tubules according to the scheme shown in figure 34. Since acid secretion in turn depends upon hydration of carbon dioxide, it is easy to suppose that an increased P_{CO_2} increases the rate of acid secretion and bicarbonate reabsorption by increasing the rate of hydration of carbonic acid and that a decreased P_{CO_2} reduces acid secretion and bicarbonate reabsorption by reducing the rate of hydration of carbonic acid. Although this is a reasonable supposition, there is no proof that it is correct.

Compensation for respiratory alkalosis is shown in figure 36. The initial hyperventilation resulting in respiratory alkalosis is represented by the arrow from point A to B. For the sake of simplicity, it is assumed that hyperventilation reduces the P_{CO_2}

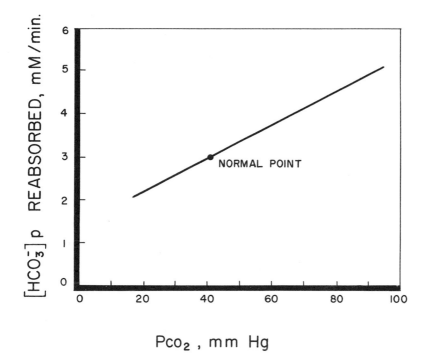

Fig. 35.   The rate of absorption of bicarbonate by the renal tubules in millimoles per minute plotted against the plasma P_{CO_2}. The rate of reabsorption of bicarbonate is calculated for a man having a constant rate of glomerular filtration of 0.125 liter per minute. Freely adapted from the data of Brazeau and Gilman, 1953, *Am. J. Physiol.* 175:33. (Reproduced by permission.)

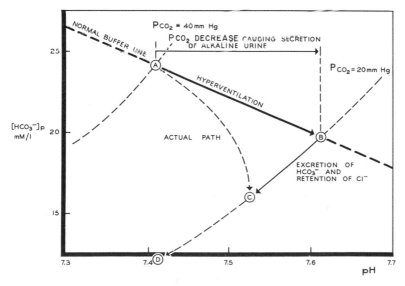

Fig. 36.   Processes occurring in the renal compensation for respiratory alkalosis.

to 20 mm Hg and keeps it at this value. Decreased P_{CO_2} moves the blood down its normal *in vivo* buffer line. Plasma pH is raised, and plasma bicarbonate concentration is reduced. The effect of low P_{CO_2} is to reduce tubular acid secretion and bicarbonate reabsorption, and an alkaline urine containing bicarbonate appears in the bladder. Excretion of bicarbonate occurs despite the fact that respiratory alkalosis has already lowered plasma bicarbonate, and excretion of bicarbonate into the urine lowers it still further. As bicarbonate is excreted in the urine, chloride is conserved, raising plasma chloride to replace lost bicarbonate. As hyperventilation continues, the blood moves down the P_{CO_2} equals 20 mm Hg isobar, as shown by the arrow from B to C. Movement in this direction involves a decrease in plasma pH or a return of plasma pH toward normal. This process may continue until point D is reached, at which the blood pH has returned to normal, and respiratory alkalosis is completely compensated. Since both processes – the initial hyperventilation and its compensation – occur together, the actual acid-base path is the one represented by the broken arrow from A to C.

Often, but not invariably, development of respiratory alkalosis is rapidly followed by an increase in the concentration of lactate and pyruvate in the blood. For some unknown reason, hyperventilation causes tissues to release lactic and pyruvic acids; this produces a base deficit. The effect on plasma pH and bicarbonate concentration is the same as that produced by renal compensation: return of pH toward normal and reduction of bicarbonate concentration. In some instances the whole compensation can be attributed to accumulation of lactate and pyruvate; in other instances compensation is effected partly by metabolic acidosis of lactic and pyruvic acid production and partly by renal compensation.

Compensation for respiratory acidosis is shown in figure 37. The initial hypoventilation resulting in respiratory acidosis is represented by the arrow from the normal point to B. Hypoventilation increases the P_{CO_2} of the blood and moves it up its *in vivo* buffer line, increasing plasma bicarbonate concentration and decreasing pH. For the sake of simplicity, it is assumed that hypoventilation raises the P_{CO_2} to 60 mm Hg and keeps it at this value. Increase in P_{CO_2} causes the kidney to secrete more acid and to reabsorb bicarbonate at a greater rate. Despite the fact that the plasma bicarbonate concentration is elevated and the rate of glomerular filtration of bicarbonate is increased, all filtered bicarbonate is reabsorbed, and acid urine containing no bicarbonate is excreted. Secretion of acid urine adds base to the blood, further raising the plasma bicarbonate concentration. Along with these changes goes increased excretion of chloride, so that

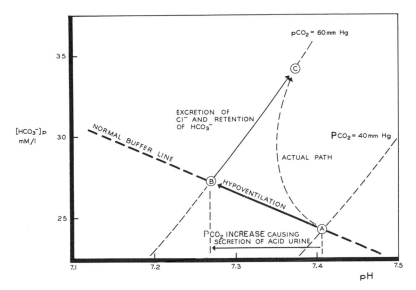

Fig. 37.   Processes occurring in the renal compensation for respiratory acidosis.

plasma chloride concentration is lowered to approximately the same extent that plasma bicarbonate concentration is raised. Compensation in the sense of a return of pH toward normal occurs at the expense of further distortion of plasma bicarbonate and chloride concentrations. If retention of bicarbonate and excretion of acid urine continue, the blood moves up the P_{CO_2} equals 60 mm Hg isobar until normal pH is reached. In this example, compensation is complete at a bicarbonate concentration of 36 millimoles per liter.

In earlier sections of this book renal responses to metabolic alkalosis and acidosis were discussed without reference to changes in renal tubular acid secretion and bicarbonate reabsorption produced by changes in P_{CO_2}. This factor was omitted for sake of simplicity of exposition, for it has little importance in these conditions. Actually, changes in P_{CO_2} do occur in respiratory compensation for metabolic alkalosis, and to some extent these P_{CO_2} changes do influence renal function. From the point of view of compensation, these changes influence renal function in the wrong direction. In metabolic alkalosis the P_{CO_2} is, if anything, elevated, and this tends to increase tubular acid secretion and bicarbonate reabsorption. However, the total renal response results in decreased acid excretion and increased bicarbonate excretion. The reason is that, despite slightly elevated tubular acid secretion caused by higher P_{CO_2}, the much greater rate of filtration of bicarbonate caused by elevated plasma bicarbonate concentration overwhelms the tubular reabsorbing mechanism, and not all bicarbonate presented to the tubules in the glomerular filtrate can be reabsorbed. Bicarbonate not reabsorbed is excreted, and plasma bicarbonate falls.

In metabolic acidosis respiratory compensation lowers the P_{CO_2}. This reduces renal tubular secretion of acid. However, plasma bicarbonate concentration falls as well, and consequently the amount of bicarbonate filtered per unit time is also reduced. Since by far the largest fraction of the acid secreted by the tubules is used to reabsorb bicarbonate, the amount of acid required for bicarbonate reabsorption is drastically reduced. Although the rate of acid secretion by the tubules is lower than normal, the concomitant decrease in the demand for acid to be used in bicarbonate reabsorption leaves more acid to be excreted in the urine as titratable acidity and ammonium. A numerical example illustrating these principles is given in table 12.

2.11. Identification of Acid-Base Status

Accurate analysis of blood gives values for plasma pH and bicarbonate concentration and for the P_{CO_2}. Each of these may be high, low, or normal, and there are nine ways in which the values may be combined (fig. 27). Whatever combination of values exists in blood is the result of physiological causes, and determination of acid-base status in itself does not reveal the cause or the route by which that status was reached.

Table 12

Renal handling of Bicarbonate and Acid in a Normal Person and in One with Partially Compensated Metabolic Acidosis. Glomerular filtration rate (GFR) is normal in both.

	Normal	*Metabolic Acidosis*
GFR, liters per day	180	180
$[HCO_3^-]_p$, milliequivalents per liter	24	12
HCO_3^- filtered, milliequivalents per day	4,320	2,160
HCO_3^- reabsorbed, milliequivalents per day	4,315	2,160
Titratable acid plus NH_4^+ in urine, milliequivalents per day	60	200
Acid secreted by renal tubules		
To reabsorb bicarbonate	4,315	2,160
To be excreted in urine	60	200
Total, milliequivalents per day	4,375	2,360

There are three ways a particular acid-base status may be attained:

1. A single physiological process, respiratory or metabolic, may be at work. Hyperventilation may cause respiratory alkalosis; or vomiting of acid gastric juice may produce metabolic alkalosis.
2. A single physiological process which produces a deviation from normal may be followed by normal physiological compensation. Respiratory alkalosis of hyperventilation may be corrected by excretion of alkaline urine, or metabolic alkalosis may be partially compensated by secondary respiratory acidosis.
3. Multiple causes, each with or without compensation, may produce a mixed status.

    The first step in the identification of acid-base status is to determine plasma pH and bicarbonate concentration and the P_{CO_2}. If two of these are known, the third can be calculated. If only one is known, the acid-base status cannot be understood. It is absolutely essential to know two facts about blood; one is not sufficient.

> Example 21.*   A patient was hyperventilating. Her plasma pH was not measured, but its total carbon dioxide content was 5 millimoles per liter.
>
> *Interpretation.*   A reasonable estimate of plasma bicarbonate concentration can be made. Within the usual range of P_{CO_2} from 20 to 60 mm Hg dissolved carbon dioxide of plasma varies between 0.6 and 1.8 millimoles per liter. In this instance the fact the patient was hyperventilating indicates that the P_{CO_2} was probably low. Therefore, plasma bicarbonate concentration lay between 4 and 5 millimoles per liter. A broken line is drawn across figure 38 representing this plasma bicarbonate concentration.
>
> If plasma pH is not known, the position of the point along the line is likewise unknown. Low plasma bicarbonate concentration might be reached by way of metabolic acidosis for which hyperventilation is a compensatory process, or it might be reached by way of respiratory alkalosis resulting from primary hyperventilation. In the latter case, there might be either the normal renal compensation consisting of excretion of base and retention of acid, or there might be an additional metabolic acidosis such as that caused by excessive production of lactic and pyruvic acids.

*Adapted from S. R. Gambino, 1965, *New Eng. J. Med.* 272:541. (Reproduced by permission.)

In either case, a base deficit accounts in part for the low plasma bicarbonate concentration.

The physician recognized the base deficit, but he assumed that it must be caused by primary metabolic acidosis. He gave the patient 225 milliequivalents of sodium bicarbonate intravenously to correct metabolic acidosis.

Hyperventilation persisted. Six hours later another blood sample was taken. Plasma pH was 7.52, and its total carbon dioxide content was 8 millimoles per liter. From these data the bicarbonate concentration was calculated to be 7.7 millimoles per liter, and the P_{CO_2} was found to be 10 mm Hg. This point is plotted as *2* in figure 38.

Interpretation. The blood had a base deficit of -13 millimoles per liter. Now an additional datum, the plasma pH, allows a decision to be made whether base deficit is the result of primary metabolic acidosis, as had originally been assumed, or whether it was secondary to primary hyperventilation. Respiratory compensation for metabolic acidosis does not overshoot the normal pH, and high pH itself inhibits respiration. Therefore, the hyperventilation must be a cause, not an effect, and its etiology must be sought elsewhere than as compensation for metabolic acidosis. Initial administration of sodium bicarbonate, based on the

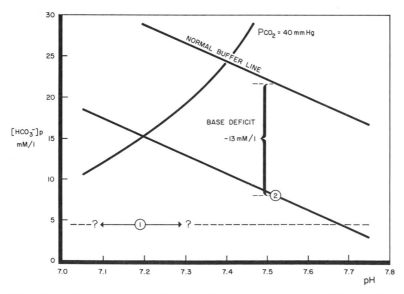

Fig. 38.   pH-Bicarbonate diagram illustrating the impossibility of distinguishing between respiratory alkalosis and metabolic acidosis if only the total carbon dioxide of the plasma is known.

assumption that hyperventilation was secondary to metabolic acidosis, was inappropriate.

Mixed acid-base disturbances cannot be understood from a study of the blood alone; each cause must be sought and its effect evaluated.

Example 22.   A patient with obstructive pulmonary disease had labored breathing. She was cyanotic and irrational. The following data were obtained on an arterial blood sample:

$$Hemoglobin\ saturation = 59\%$$
$$P_{CO_2} = 67\ mm\ Hg$$
$$Plasma\ pH = 7.40$$
$$[HCO_3^-]_p = 40\ mM/l$$
$$[Na^+]_p = 134\ mM/l$$
$$[K^+]_p = 5.3\ mM/l$$
$$[Cl^-]_p = 81\ mM/l.$$

Bicarbonate and pH are plotted as point *1* in figure 39.

When the patient was given oxygen to breathe, she lapsed into a deep coma.

Interpretation.   Known obstructive pulmonary disease, low arterial hemoglobin saturation, and high P_{CO_2} all indicate that a primary cause of acid-base disturbance was the patient's inability to maintain adequate ventilation of the

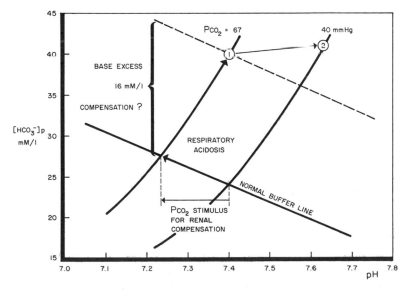

Fig. 39.   The acid-base status of a patient with respiratory acidosis partially compensated but mixed with metabolic alkalosis.

lungs. Consequently, the P_{O_2} of the alveolar air fell, and the P_{CO_2} rose to 67 mm Hg. This increase in P_{CO_2} titrated the blood up its *in vivo* buffer line in the direction of low pH and high bicarbonate concentration as shown in figure 39, by the arrow representing respiratory acidosis. Had there been no other change in acid-base status, the pH of the blood would have fallen to 7.23, and the bicarbonate concentration of plasma would have risen to 27 millimoles per liter.

Increased P_{CO_2} could cause renal compensation by increasing renal tubular acid secretion and bicarbonate reabsorption. Acid urine containing a high concentration of chloride was probably excreted, and chloride removed from blood was replaced by bicarbonate ions. A base excess occurred, and plasma bicarbonate rose to 40 millimoles per liter. Plasma pH returned to the normal value of 7.40, and the respiratory acidosis was completely compensated.

In figure 39 a broken line through point *1* is drawn parallel to the normal buffer line. If this buffer line can be taken to represent the patient's actual *in vivo* buffer line, the vertical distance between the two lines represents a base excess of 16 millimoles per liter.

A patient with arterial hemoglobin saturation of 59 per cent and a P_{CO_2} of 67 mm Hg has an arterial P_{O_2} of about 40 mm Hg. This low partial pressure of oxygen strongly stimulated the patient's peripheral chemoreceptors, and the resulting reflex drove a respiratory center which was otherwise depressed by hypoxia. It is likely that the patient's central nervous system was also depressed by the acutely elevated P_{CO_2}. Administration of oxygen relieved hypoxia, but it also removed the reflex drive from the chemoreceptors. Ventilation of the lungs decreased, and there was a still greater retention of carbon dioxide. The patient's coma following oxygen administration could be attributed to carbon dioxide narcosis.

Further clinical course. Controlled respiration was begun to reduce the P_{CO_2}. Hemoglobin in arterial blood became 96 per cent saturated, and alveolar P_{CO_2} fell to 40 mm Hg. These are entirely satisfactory values, and one would expect that the acid-base status of the patient would return to normal. With the P_{CO_2} at 40 mm Hg the stimulus for renal compensation is gone, and secretion of acid by the renal tubules should fall. On account of its high concentration in plasma, the rate of glomerular filtration of bicarbonate would exceed the rate of secretion of acid by the tubules. Some bicarbonate would escape into the urine; chloride would be conserved; and the base excess would be eliminated.

Instead of improving, the patient's condition deteriorated over the next few days. An arterial blood sample had a plasma pH of 7.63, and the plasma bicarbonate concentration was 41 millimoles per liter. This is plotted as point *2* in figure 39. The patient continued to excrete acid urine instead of the alkaline urine which would correct her metabolic alkalosis.

Interpretation.   Correction of respiratory deficiencies unveiled a profound post-hypercapnic metabolic alkalosis which was no longer appropriate compensation for respiratory acidosis. Another cause was sought. The failure of the kidney to get rid of excess plasma bicarbonate was found to be the result of chloride deficiency. Tubular reabsorption of sodium is accompanied by passive reabsorption of an equal amount of anion. With plasma chloride, and hence tubular urine chloride, concentration low, most of the bicarbonate present in renal tubular fluid was reabsorbed with the sodium. Potassium deficiency may also have contributed to the kidney's inability to excrete alkaline urine, for the condition was rapidly corrected by administration of potassium chloride.

2.12. Clinical Example: Metabolic Acidosis*

The patient, a twenty-three-year-old woman, was admitted to the hospital on January 29, 1923. Diabetes mellitus had been recognized six weeks before admission. On admission, the patient was hyperpneic and dehydrated, and there were high concentrations of acetoacetic acid and glucose in her urine. In addition, she was suffering from numerous furuncles. At the time of admission, supplies of insulin were limited, and the patient was placed on a strict dietetic regimen. By February 7, when blood sample no. 1 was taken, signs of uncontrolled diabetes had disappeared. At this time the patient was started on a daily dose of 20 units of insulin.

Three days later a carbuncle developed, and the patient's diabetes became completely uncontrolled. On February 13, she was comatose and severely hyperpneic, and at this time blood sample no. 2 was taken. The dose of insulin was doubled, and

*The data in this section are from G. E. Cullen and L. Jonas, 1923, *J. Biol. Chem.* 57:541. (Reproduced by permission.)

the patient rapidly improved. Blood samples nos. 3 and 4 were taken on the two following days.

The data obtained on the blood samples are given in table 13, and the points are plotted on pH-bicarbonate diagrams in figures 40 and 41.

Table 13

No.	Date	pH	Plasma $[HCO_3^-]_p$	P_{CO_2}
1	2/7/23	7.39	19.0 mM/l	33 mm Hg
2	2/13/23	7.02	6.2	24
3	2/14/23	7.39	15.2	28
4	2/15/23	7.42	25.2	40

At the time the first blood sample was taken, the patient was in almost completely compensated metabolic acidosis. The acidosis was not severe, for, as shown in figure 41, the patient's base deficit was only −5 millimoles per liter. Had there been no respiratory compensation, the pH of the plasma would have been 7.33. Hyperventilation reduced the P_{CO_2} to 33 mm Hg and raised the pH to 7.39.

By the time blood sample no. 2 was taken, the patient had developed severe metabolic acidosis, and, if there had been no respiratory compensation, the pH of the plasma would have been 6.93. The patient's hyperventilation had reduced the P_{CO_2} to 24 mm Hg. Even this lowering of the P_{CO_2} was insufficient to raise the pH higher than 7.02.

Between samples 1 and 2 a base deficit of 22 millimoles per liter had accumulated. This means that 22 millimoles of strong acid had been added to each liter of blood. However, the increase in plasma hydrogen ion concentration was only from 0.000,041 millimoles per liter (pH 7.39) to 0.000,095 millimoles per liter (pH 7.02). The increase in $[H^+]_p$, 0.000,054 millimoles per liter, was less than 0.000,3% of the amount of strong acid added to each liter of blood.

The most vigorous hyperventilation would have been ineffective in restoring the pH to normal. In figure 41 a broken line is drawn through point 2 parallel to the normal buffer line. Increased hyperventilation moves blood down this line to the right. If the patient had been able to hyperventilate enough to reduce her P_{CO_2} to the very low value of 5 mm Hg and her plasma bicarbonate to 2 millimoles per liter, the pH would have risen only to 7.20.

The increased dose of insulin given after occurrence of severe metabolic acidosis permitted removal of acetoacetate

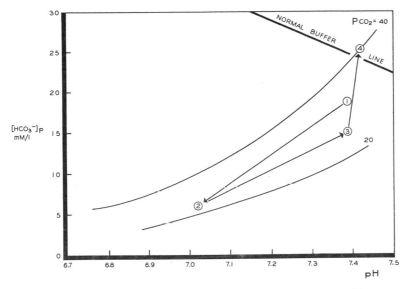

Fig. 40.   The acid-base pathway of a patient with diabetes mellitus.

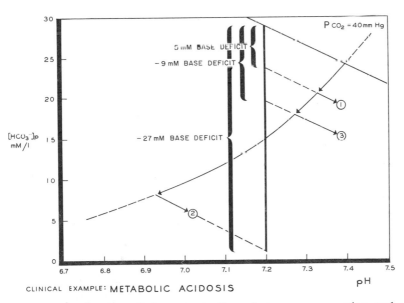

CLINICAL EXAMPLE: METABOLIC ACIDOSIS

Fig. 41.   Estimation of the extent of respiratory compensation and base deficit in a patient with diabetes mellitus.

from the blood and correction of the base deficit. By the time blood sample no. 3 was taken, the base deficit was −9 millimoles per liter. Hyperventilation, by reducing the P_{CO_2} to 28 mm Hg, raised plasma pH to 7.39. The next day when blood sample no. 4 was taken, the deficit had been completely corrected.

2.13. Clinical Example: Respiratory Alkalosis*

The patient, a five-year-old girl, was admitted to the hospital on February 13, 1942, with a history of joint pains of short duration. Her heart rate was 120, and the heart was slightly enlarged. Her temperature was 101.4°F, and her respiratory rate was 32. Blood sample no. 1 was taken. A diagnosis of rheumatic fever was made.

On the day after admission the patient was started on sodium salicylate therapy, the dosage being 0.23 g per kg per day. During the next six days the patient was nauseated, listless, and drowsy; and she vomited eight times. On the day after beginning of medication the patient's respiratory rate rose to 35, but thereafter it subsided to 25 or less.

Blood sample no. 2 was taken two days, and blood sample no. 3 six days after beginning of medication. Urine samples were obtained on the same days. The salicylate therapy was discontinued after six days, and the patient's vomiting stopped. Blood sample no. 4 was taken three days later.

Table 14

No.	Date	pH	Plasma $[HCO_3^-]_p$	P_{CO_2}	Urine pH
1	2/13/42	7.43	24.4 mM/l	39 mm Hg	
2	2/16/42	7.57	23.7	27	6.0
3	2/20/42	7.41	13.5	22	6.0
4	2/23/42	7.41	23.4	39	5.0

Analytical figures on true plasma and urine are given in table 14, and the points are plotted on pH-bicarbonate diagrams in figures 42, 43, 44, and 45.

At the time the patient was admitted to the hospital the acid-base pattern of her blood was normal. Despite the rather high respiratory rate of 32, the P_{CO_2} of the blood was 39 mm Hg, and the minute volume of alveolar ventilation must have been normal.

Blood sample no. 2, taken two days after the beginning of medication, had a high pH and a low P_{CO_2}. Bicarbonate concentration was normal. P_{CO_2} of the blood can be decreased only by hyperventilation, and the patient's hyperventilation was caused by sodium salicylate which stimulates respiration. Although the

*The data in this section are from G. M. Guest, S. Rapoport, and C. Roscoe, 1945. *J. Clin. Invest.* 24:770, together with additional data supplied by the authors. (Reproduced by permission.)

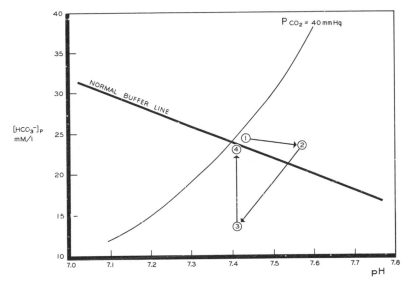

Fig. 42. The acid-base pathway of a patient on sodium salicylate medication.

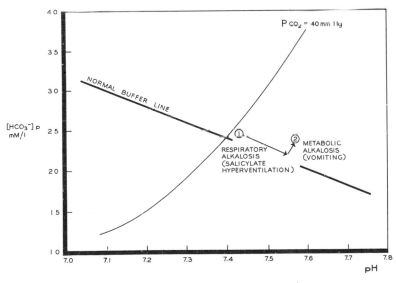

Fig. 43. The initial development of respiratory alkalosis and metabolic alkalosis in a patient on sodium salicylate medication.

patient's respiratory rate was not greatly elevated over that observed at the time of admission, the minute volume of alveolar ventilation must have been far above normal. This hyperventilation, by decreasing the P_{CO_2} of the blood, caused the blood to be titrated along its normal buffer line in the direction of high

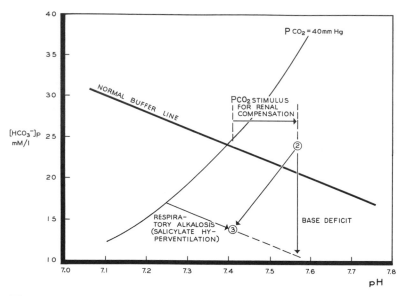

Fig. 44.   Accumulation of base deficit in a patient on sodium salicylate medication.

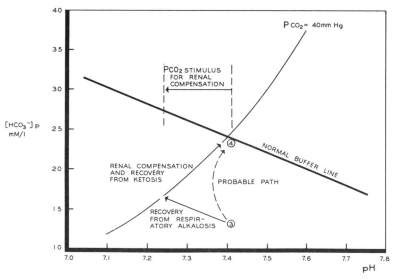

Fig. 45.   The return to normal after withdrawal of sodium salicylate medication.

pH and low bicarbonate concentration, as shown by the arrow from point *1* in figure 43. However, point *2* is not on the normal buffer line but lies above it. Bicarbonate concentration of the plasma should have been below normal if hyperventilation were the only cause of acid-base disturbance. Therefore, the acid-base status at point 2 must be a mixed one. The likely reason for the component of metabolic alkalosis is loss of acid by vomiting.

Blood sample no. 3, taken after six days of sodium salicylate medication, had a normal pH and low P_{CO_2} and bicarbonate concentration. The fact that the P_{CO_2} was only 22 mm Hg means that the patient had a very large minute volume of alveolar ventilation and that respiratory alkalosis was an important factor in the acid-base pattern. The low concentration of bicarbonate means that base deficit had developed. Point *3* lies below the normal buffer line and shows that the base deficit was about −12 millimoles per liter of blood. Combination of base deficit and respiratory alkalosis brought the pH of plasma to its normal value.

There are four general reasons for occurrence of base deficit in this instance:

1. Renal compensation, excretion of base, and retention of acid may have occurred. The fact that the pH of the urine was never alkaline suggests that renal compensation was a minor component of the base deficit.
2. Addition of lactic and pyruvic acids to the blood often follows respiratory alkalosis of any origin. Because the blood was not analyzed for lactate and pyruvate, this possible cause of base deficit cannot be evaluated.
3. Salicylate poisoning in itself appears to cause metabolic acidosis, the means by which it does so being unknown.
4. Undernutrition, particularly in children, often causes ketosis. At the time blood sample no. 3 was taken the patient had been nauseated and vomiting for six days. Under these circumstances it is unlikely that a child of five would have eaten enough to keep up her carbohydrate stores. Once the stores had been depleted, her body fat would have served as a major source of energy. Increased fat metabolism beyond a certain point results in an increase of ketone acids in the blood and consequently in metabolic acidosis.

When the salicylate medication was withdrawn, the acid-base pattern rapidly returned to normal. The stimulus for hyperventilation was removed, and the P_{CO_2} rose. Renal tubular reabsorption of bicarbonate increased, and plasma bicarbonate rose. With disappearance of nausea and vomiting and excretion of remaining salicylate, metabolic acidosis disappeared.

Other Ways of Looking at the Problem

3.1. The pH-Log P_{CO_2} Diagram

In the pH-log P_{CO_2} diagram the logarithms of the partial pressures of carbon dioxide measured in mm Hg are plotted as ordinates against the pH values as abscissas. It is convenient to use one-cycle semi-logarithmic paper in which the pH values are on a linear horizontal scale and the P_{CO_2} values are on a logarithmic vertical scale. When this paper is used, the P_{CO_2} values can be plotted without having to look up their logarithms. Horizontal lines on the graph are P_{CO_2} isobars, and the P_{CO_2} equals 40 mm Hg isobar is drawn in figure 46.

The pH, P_{CO_2}, and bicarbonate concentration of plasma are related by equation 50 (sec. 1.13):

$$pH = pK + \log \frac{[HCO_3^-]_p}{a \, P_{CO_2}}.$$

Rearranging and changing signs, we have

$$\log P_{CO_2} = pK - \log a + \log [HCO_3^-]_p - pH. \tag{78}$$

For any given constant value of bicarbonate concentration log $[HCO_3^-]_p$ is also constant. Equation (78) then reduces to that for a straight line

$$y = A + bx. \tag{79}$$

In this equation

$$y = \log P_{CO_2},$$
$$A = (pK - \log a + \log [HCO_3^-]_p),$$
$$b = -1$$

and $x = pH$.

Therefore, when $[HCO_3^-]_p$ is constant at any particular value, there is a line on the graph having a slope of -1 which is the locus of all combinations of pH and log P_{CO_2} satisfying equations (50) and (78). This line of constant bicarbonate concentration is, for the pH-log P_{CO_2} diagram, the equivalent of the P_{CO_2} isobar on the pH-bicarbonate diagram.

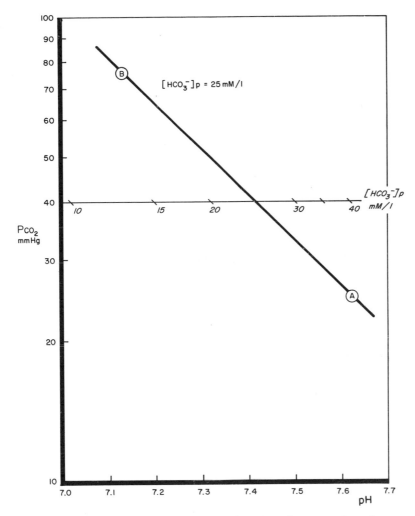

Fig. 46.   The pH-log P_{CO_2} diagram. The pH values are plotted on a
horizontal linear scale, and the P_{CO_2} values are plotted on a vertical
logarithmic scale. Horizontal lines are P_{CO_2} isobars, and that for 40 mm
Hg is marked. The line for constant bicarbonate concentration of 25
millimoles per liter is drawn, and the 40 mm Hg isobar is marked at
intercepts of similar constant bicarbonate lines of 10, 15, 20, 30, 35,
and 40 millimoles per liter.

Example 23.   Calculate the constant bicarbonate line for
the value $[HCO_3^-]_p$ equals 25 millimoles per liter.

First Step.   Equation (50) with appropriate numerical
values for plasma at 37°C is

$$pH = 6.10 + \log \frac{[HCO_3^-]_p}{0.0301\, P_{CO_2}}.$$

To calculate one point when $[HCO_3^-]_p$ is 25 millimoles per liter, the value of 25 mm Hg is assumed for the P_{CO_2}. Then the equation becomes

$$pH = 6.10 + \log \frac{25}{0.0301\ (25)},$$
$$= 6.10 + \log\ (33.3),$$
$$= 6.10 + 1.52,$$
$$= 7.62.$$

This point is plotted as point A in figure 46.

Second Step. A line having a slope of -1 cuts ordinates and abscissas at an angle of 45°, and such a line could be drawn through point A. Alternatively, the line can be fixed by calculating one other point. Assuming that $P_{CO_2} = 75$ mm Hg, the calculated pH is 7.14. This is plotted as point B, and a straight line is drawn through the two points.

Intercepts of the constant bicarbonate concentration lines with the isobar representing P_{CO_2} equals 40 mm Hg can be calculated in a similar fashion.

Example 24. Calculate the intercepts with the isobar representing P_{CO_2} equals 40 mm Hg of the lines for constant bicarbonate concentrations of 10, 15, 20, 30, 35, and 40 millimoles per liter.

First Step. For $[HCO_3^-]_p = 10$ millimoles per liter and $P_{CO_2} = 40$ mm Hg equation (50) becomes

$$pH = 6.10 + \log \frac{10}{0.0301\ (40)},$$
$$= 6.10 + \log 8.30,$$
$$= 6.10 + 0.92,$$
$$= 7.02.$$

This point is marked on the isobar in figure 46 with a short line having a slope of -1.

Second Step. The calculation is repeated for the other bicarbonate concentrations:

$[HCO_3^-]_p$	pH
10	7.02
15	7.20
20	7.32
25	7.42
30	7.50
35	7.56
40	7.62

These are all marked on the isobar.

Table 15

| | True Plasma No. | | | |
	1	2	3	4
P_{CO_2}, mm Hg	85.1	46.5	33.3	23.3
pH	7.16	7.35	7.46	7.57

The data establishing the normal *in vitro* buffer line of the blood of A.V.B., taken from table 5, are given in table 15, and the line is plotted in figure 47.

The slope of the normal *in vitro* buffer line of true plasma is a function of the hemoglobin content of the blood. To show the extremes, the line for separated plasma, that is, for blood having zero hemoglobin concentration, is plotted in figure 47, from the data contained in table 4. Likewise, the line for blood having 20 grams of hemoglobin per 100 ml, or 12 millimoles per liter, from figure 17 is plotted in figure 47.

Normal *in vivo* acid-base paths described in tables 9, 10, and 11 are plotted in figure 48 from the data collected in table 16.

Normal ranges of values given in section 2.5 are

$$P_{CO_2} \text{ (mm Hg)} - 35\text{–}48$$
$$pH - 7.35\text{–}7.45$$
$$[HCO_3^-]_p \text{ (mM per liter)} - 23\text{–}28.$$

These form the boundaries of the hexagon in figure 49. The areas in the figure are labeled:

1. Any condition represented by a point falling within the area above the normal P_{CO_2} isobar has a component of *respiratory acidosis*.

Table 16

Point	Ventilation	Base Status	P_{CO_2}	pH	$[HCO_3^-]_p$
A	normal	normal	39 mm Hg	7.42	24.8 mM/l
B	low P_{CO_2}	normal	20	7.62	19.9
C	high P_{CO_2}	normal	47	7.36	26.6
D	normal	base excess	42	7.52	33.7
E	low P_{CO_2}	base excess	26	7.67	29.6
F	high P_{CO_2}	base excess	57	7.42	36.5
G	normal	base deficit	37	7.35	19.8
H	low P_{CO_2}	base deficit	17	7.54	14.4
I	high P_{CO_2}	base deficit	50	7.27	22.5

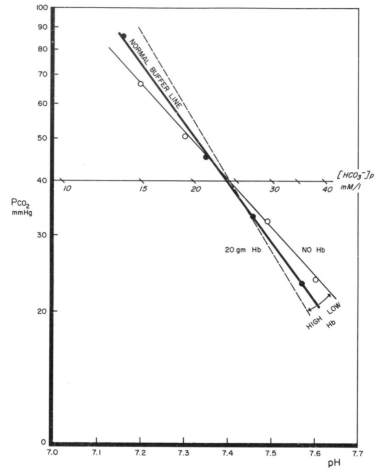

Fig. 47. The normal *in vitro* buffer line of A.V.B.'s blood plotted on the pH-log P_{CO_2} diagram together with the lines for separated plasma (no Hb) and for blood containing 20 grams of hemoglobin per 100 milliliters.

2. Any condition represented by a point falling within the area below the normal P_{CO_2} isobar has a component of *respiratory alkalosis*.

3. Any condition represented by a point falling within the area to the left of the normal pH has *low* pH.

4. Any condition represented by a point falling within the area to the right of the normal pH has *high* pH.

5. Any condition represented by a point falling within the area above and to the right of the normal buffer line has a component of *base excess*.

6. Any condition represented by a point falling within the area below and to the left of the normal buffer line has a component of *base deficit*.

The areas bounded by the 95% confidence lines plotted in figures 18 and 28 are very nearly coincident with the area bounded by the two straight lines on each side of the *in vitro* normal buffer line in figure 49. The whole-body buffer line plotted from the coordinates given in table 7 falls within the area bounded by these two straight lines. Although figure 49 and all commercially available charts similar to it are based on blood *in vitro*, they can be considered to approximate the behavior of blood *in vivo*.

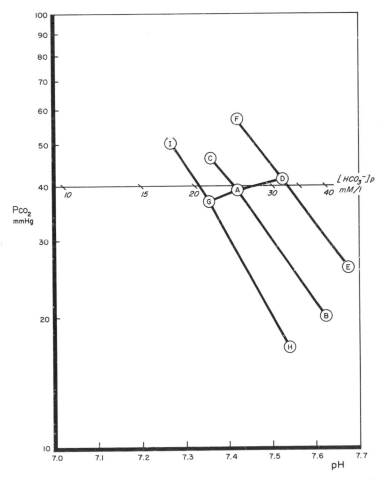

Fig. 48. The nine combinations of P_{CO_2}, pH, and base excess or deficit plotted on a pH-log P_{CO_2} diagram. See table 16 for identification of the points.

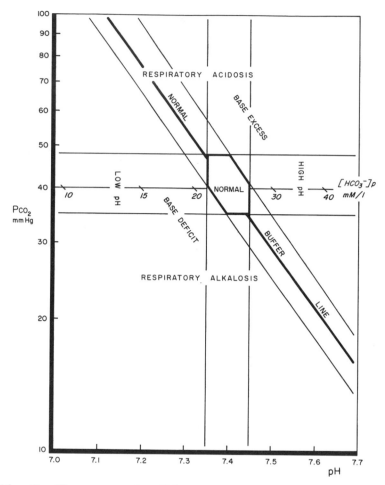

Fig. 49.   The area on the pH-log P_{CO_2} diagram within which normal values fall.

The pH-log P_{CO_2} diagram as it stands in figure 49 contains no more information than does the pH-bicarbonate diagram; it can help the physician identify the patient's acid-base status, but it cannot reveal the underlying causes of abnormal patterns, nor can it point the way to appropriate treatment.

Clinical Example.   The data obtained on the blood of the patient with respiratory acidosis described in section 2.11 are given:

Initial untreated respiratory acidosis (point *1*):

pH = 7.40

P_{CO_2} = 67 mm Hg.

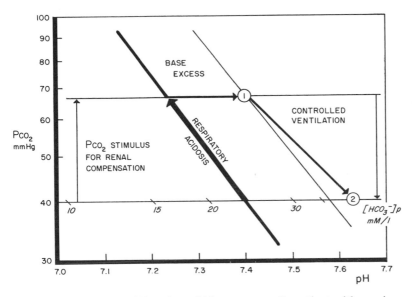

Fig. 50.   Points describing the acid-base status of a patient with respiratory acidosis (*1*) treated with controlled respiration (*2*). See section
2.11 for details.

After controlled respiration (point *2*):

$$pH = 7.63$$
$$P_{CO_2} = 40 \text{ mm Hg.}$$

The points are plotted on a pH-log P_{CO_2} diagram in figure 50.

3.2. Determination of Acid-Base Status of Blood by Equilibration with Gas Mixtures of Known P_{CO_2}

Apparatus available in modern clinical laboratories allows determination of the acid-base status of a small sample of blood by
means of three measurements:

1. The pH of the blood sample as drawn from the subject is measured. This is, in effect, the pH of the plasma.*

---

*When pH determinations are made on whole blood, the pH value ob-
tained is that of plasma very slightly modified by presence of erythrocytes. This is
not because erythrocytes affect the pH of plasma, but it is because they affect its
measurement. Precipitation of erythrocyte proteins at the liquid junction between
blood and the saturated potassium chloride solution making electrical connection
with the reference electrode affects the junction potential. The true pH of plasma
is about 0.01 pH unit higher than that measured on whole blood, but the difference
is usually ignored. See Siggaard-Andersen, 1961, *Scand. J. Clin. Lab. Invest.*
13:205.

2. A sample of the blood is equilibrated with a gas mixture saturated with water vapor and containing a low, precisely known percentage of carbon dioxide and enough oxygen to saturate the hemoglobin. The pH of the equilibrated blood is measured, and the barometer is read.
3. Another sample of the blood is equilibrated with a gas mixture saturated with water vapor and containing a high, precisely known percentage of carbon dioxide and enough oxygen to saturate the hemoglobin. The pH of the equilibrated blood is measured.

The apparatus is built so that all determinations can be made at normal body temperature, 37°C, or at any temperature corresponding to that of the patient under study.

From these data the P_{CO_2} and the bicarbonate concentration of the plasma of the original blood can be calculated. A pH-log P_{CO_2} diagram is particularly useful for this purpose.

Example 25. The pH of a sample of blood, measured at 37°C, is 7.23. Determine the P_{CO_2} and the plasma bicarbonate concentration.

First Step. The blood is equilibrated at 37°C with a gas mixture containing 2.66% carbon dioxide and about 25% oxygen. The prevailing barometric pressure (P_B) is 743 mm Hg. The pH of the blood, measured at 37°C, is found to be 7.32.

The P_{CO_2} of the equilibrated blood is calculated.

$$P_{CO_2} = (P_B - P_{H_2O})(\% \ CO_2)/100.$$

The vapor pressure of water at 37°C is 47 mm Hg.

$$P_{CO_2} = (743 - 47)(2.66)/100,$$
$$= 18.5 \ \text{mm Hg}.$$

The point representing this equilibrated sample is plotted at A on the pH-log P_{CO_2} diagram in figure 51.

Second Step. Another sample of the blood is equilibrated at 37°C with a gas mixture containing 9.33% carbon dioxide and about 25% oxygen. The P_{CO_2} is 65.0 mm Hg. The pH is found to be 7.03. The point is plotted at B in figure 51.

Points A and B are two points on the experimentally determined *in vitro* buffer line of this particular sample of blood. Since the buffer line is essentially linear over this pH range, a straight line drawn through the two points gives the actual *in vitro* buffer line of the blood.

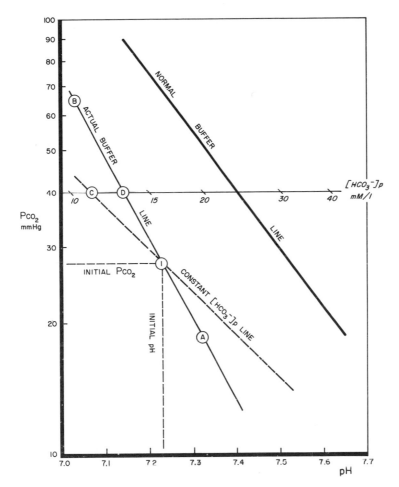

Fig. 51.   Determination of the P_{CO_2} and plasma bicarbonate concentration of a blood sample by measurement of its pH before and after equilibration with gas mixtures having known P_{CO_2}'s.

Third Step.   The pH of the original sample was 7.23. A vertical line drawn at the pH intersects the buffer line at point *1*. The horizontal P_{CO_2} isobar passing through this point is that of 27.5 mm Hg. This is the P_{CO_2} of the original blood sample.

Fourth Step.   The plasma bicarbonate concentration of the original sample can be determined in two ways.
     Substitution of the known values in the equation

$$pH = pK + \log \frac{[\text{HCO}_3^-]_p}{0.0301\, P_{CO_2}}$$

gives

$$7.23 = 6.10 + \log \frac{[HCO_3^-]_p}{0.0301 \ (27.5)},$$

$$1.13 = \log \frac{[HCO_3^-]_p}{0.828},$$

antilog $1.13 = [HCO_3^-]_p/(0.828) = 13.5,$

$$[HCO_3^-]_p = 13.5 \ (0.828),$$
$$= 11.2 \ mM/l.$$

Alternatively, the line of constant bicarbonate concentration on which point *1* lies can be determined by drawing a line having a slope of −1 through point *1* and by reading its intercept on the P_{CO_2} equals 40 mm Hg isobar. Such a line gives an intercept (point *C*) at about 11 mM/l.

The procedure of equilibrating the blood after it is drawn from the subject produces an additional datum—the slope of the *in vitro* buffer line of the blood. This slope is chiefly a function of the concentration of hemoglobin. The procedure does not give the slope of the buffer line *in vivo*, which in addition to being affected by the concentration of hemoglobin is also a function of the volume of interstitual fluid and of body buffering. The reader must remember the description of the differences between the slope of the buffer line measured *in vitro* and the actual slope of the buffer line *in vivo* contained in section 2.1.

3.3. Quantitation of the Metabolic Component: The Carbon Dioxide Combining Power

At the beginning of the twentieth century when the principles of acid-base chemistry were being developed and applied to clinical problems the only method generally available was one for measuring the total carbon dioxide content of plasma. Ability to make accurate measurement of pH was confined to a few adepts in the best research laboratories. The acid-base status of a patient is the resultant of at least two processes, one respiratory and the other metabolic; and two measurements are required to determine the contribution of each. Physiological chemists thought that if the effect of one—the respiratory component—could be eliminated, a single measurement of plasma carbon dioxide content could allow evaluation of the other—the metabolic component. To meet this need Van Slyke and Cullen (1917, *J. Biol. Chem.* 30:289) invented the concept of *carbon dioxide combining power.*

Determination of carbon dioxide combining power has long been superceded by far better methods in good laboratories, but because its determination is still practiced in some hospitals, the student must understand what information it affords.

For measurement of carbon dioxide combining power,

plasma is separated from erythrocytes. It is then equilibrated at body temperature with a gas mixture having a P_{CO_2} of 40 mm Hg, and after equilibration it is analyzed for its total carbon dioxide content. Because the plasma had been equilibrated with carbon dioxide at 40 mm Hg just before analysis, its concentration of dissolved carbon dioxide is 1.2 millimoles per liter. Subtraction of 1.2 from the total carbon dioxide content of plasma expressed in millimoles per liter gives the bicarbonate concentration of the sample. This is the carbon dioxide combining power.

Equilibration of the plasma with a gas mixture having a P_{CO_2} of 40 mm Hg, the normal value, was designed to eliminate the respiratory component by restoring the sample analyzed to the normal P_{CO_2}. Then the difference between the observed bicarbonate concentration and the normal bicarbonate concentration of 23 to 28 millimoles per liter is a measure of the metabolic component. If the carbon dioxide combining power is greater than 28 millimoles per liter, at least one component of the patient's acid-base pattern is base excess, and the magnitude of the difference is a rough measure of the severity of the base excess. If the carbon dioxide combining power is lower than 23 millimoles per liter, there is some degree of base deficit present, and the magnitude of the difference allows an estimate of its severity.

Determination of carbon dioxide combining power contains, in addition to its liability to analytical error, a theoretical error which robs it of some of its quantitative accuracy as a measure of the metabolic component. Plasma is separated from erythrocytes before it is equilibrated with the gas mixture having a P_{CO_2} of 40 mm Hg. Consequently, the plasma is separated plasma, and the slope of the line it follows during equilibration is that of the flat buffer line of separated plasma.

Results of equilibrating separated plasma are shown in figure 52. The carbon dioxide combining power might be found to be that represented by point *A*. Base excess is present, and the acid-base pattern contains a component of metabolic alkalosis. The plasma might originally have had a low P_{CO_2} as shown by the right-hand question mark, or it might have had a high P_{CO_2} as shown by the left-hand question mark. In either case, during equilibration it moved along the buffer line of separated plasma to reach point *A*.

The difference between equilibrating separated and true plasma is shown at the bottom of figure 52. Suppose that the sample of blood taken is at point *B*. If the blood is equilibrated before plasma is removed, the blood will follow the slope of the normal buffer line of true plasma during equilibration, and the carbon dioxide combining power will be found to be that represented by point *C*. If plasma is equilibrated after removal of erythrocytes,

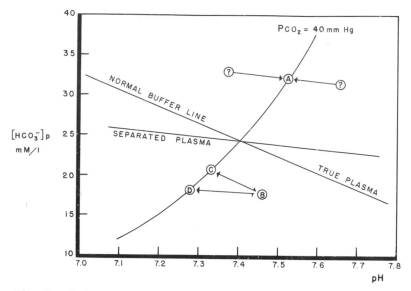

Fig. 52. Points representing disturbances in acid-base pattern of the blood to illustrate the meaning of carbon dioxide combining power measured after equilibration of separated plasma with a gas mixture having a P_{CO_2} of 40 mm Hg. Point C gives the standard bicarbonate of the blood originally at point B.

the plasma will follow the slope of the buffer line of separated plasma during equilibration, and the carbon dioxide combining power will be found to be that represented by point D. The points differ by several millimoles per liter, and the error is in point D because separated plasma was used.

Measurement of carbon dioxide combining power gives an approximate estimate of base excess or base deficit, and it gives no information whatever about the respiratory component of the acid-base pattern.

3.4. Quantitation of the Metabolic Component: Standard Bicarbonate*

Standard bicarbonate is the bicarbonate concentration in millimoles per liter of plasma of whole blood which has been equilibrated at 37°C with a gas mixture having a P_{CO_2} of 40 mm Hg.

The procedure of bringing the whole blood sample to a P_{CO_2} of 40 mm Hg before the bicarbonate concentration of plasma is determined avoids the error which equilibration of separated plasma introduces into the carbon dioxide combining power.

*Standard bicarbonate was defined by K. Jorgensen and P. Astrup, 1957, *Scand. J. Clin. Lab. Invest.* 9:122.

Point C in figure 52 is the standard bicarbonate of the blood whose original status is represented by point B. The superiority of standard bicarbonate over carbon dioxide combining power is that determination of standard bicarbonate involves the use of the true buffer line of the blood.

It is not necessary to bring the blood to a P_{CO_2} of 40 mm Hg in order to estimate the standard bicarbonate. Standard bicarbonate can be determined if the blood is equilibrated with two gas mixtures having known P_{CO_2}'s. A sample of blood is drawn, and its pH is measured at 37°C. Fractions of the blood are then equilibrated at 37°C with two gas mixtures, one having a known low P_{CO_2} and the other having a known high P_{CO_2}. The pH of each equilibrated sample is then measured. This is the procedure described in example 25 whose results were plotted in figure 51. Measurement of pH at two known P_{CO_2}'s establishes the *in vitro* buffer line of the blood being studied. The point at which this buffer line crosses the P_{CO_2} equals 40 mm Hg isobar is the standard bicarbonate.

Table 17

Point	P_{CO_2}	pH	$[HCO_3^-]_p$
1	27.5 mm Hg	7.23	11.2 mM/l
A	18.5	7.32	9.2
B	65.0	7.03	16.7

Example 26. Determine the standard bicarbonate from the data in example 25.

The data together with the calculated bicarbonate concentrations are given in table 17. The points are plotted on a pH-bicarbonate diagram in figure 53.

The *in vitro* buffer line of the blood connects points A and B, and it crosses the P_{CO_2} equals 40 mm Hg isobar at a bicarbonate concentration of 13.2 millimoles per liter. This is the standard bicarbonate.

A line is drawn through point I parallel to the buffer line of separated plasma. It intersects the P_{CO_2} equals 40 mm Hg isobar at 12.2 millimoles per liter. This is the carbon dioxide combining power, and it differs from the standard bicarbonate by 1.0 millimole per liter.

Standard bicarbonate can be determined by means of a pH-log P_{CO_2} diagram. The points in example 25 are plotted in figure 51. The buffer line crosses the horizontal P_{CO_2} equals 40 mm Hg isobar at a pH of 7.14 as shown by point D. The bicar-

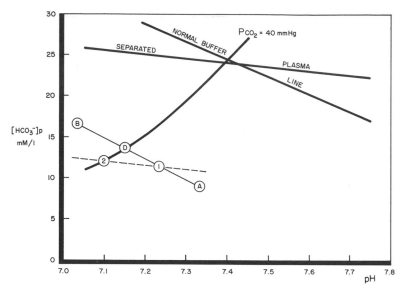

Fig. 53. The data from example 25 plotted on a pH-bicarbonate diagram to show the determination of standard bicarbonate and carbon
dioxide combining power (point 2).

bonate concentration, determined either by calculation or by
reading the bicarbonate scale already marked off on the isobar
gives a standard bicarbonate of 13.2 millimoles per liter.

The bicarbonate concentration corresponding to the point
at which the buffer line of any blood sample crosses the P_{CO_2}
equals 40 mm Hg isobar on a pH-log P_{CO_2} diagram is the standard
bicarbonate of that sample. Therefore, the standard bicarbonate
can be read from the scale marked on the isobar.

Standard bicarbonate measures the metabolic component,
and it achieves what was intended by the carbon dioxide combining power: it is a measure of the acid-base status which elimi-
nates the contribution of the respiratory component. In measuring
the metabolic component, it does not tell whether the deviation
from normal is a cause or an effect or whether the deviation is the
result of a single process or a mixture of processes.

3.5. Quantitation of the Metabolic Component:
The Base Excess Scale*

Base excess is the base concentration expressed in milliequivalents per liter measured by titration of a particular sample of blood

*Material in this section is taken from O. Siggaard-Andersen and K. Engel,
1960, *Scand. J. Clin. Lab. Invest.* 12:177, and from O. Siggaard-Andersen, 1963,
Scand. J. Clin. Lab. Invest., vol. 15, suppl. 70. (Reproduced by permission.)

at 37°C with strong acid (HCl or its equivalent) to pH 7.40 at a P_{CO_2} of 40 mm Hg. For negative values the titration is carried out with strong base (NaOH or its equivalent) to the same end point. Base excess measures both standard bicarbonate and the base taken up or given off by blood buffers when the blood moves from its normal point to its new condition. Therefore, the numerical value of base excess depends upon the *in vitro* buffer value of the blood being studied, and the *in vitro* buffer value of the blood in turn depends upon its hemoglobin content.

The argument developed in sections 2.2 and 2.3 shows that if the buffer line of blood is parallel to the normal buffer line on the pH-bicarbonate diagram, base excess can be measured by the vertical distance between the two lines. If the lines are not parallel, the measurement is to some extent in error. The error can be reduced by adding to the pH-bicarbonate diagram a base excess scale constructed from experimental observations which eliminate the effect of variations in buffer value of the blood.

A sample of blood having no base excess or deficit is obtained from a normal human subject. Part of the blood is centrifuged to provide a supply of separated plasma, and the rest of the whole blood sample is saved. Two gas mixtures are prepared, one having a precisely known high P_{CO_2} and the other having a precisely known low P_{CO_2}. In the example given in table 18 these P_{CO_2}'s are 66.0 and 28.7 mm Hg.

Base excess of exactly known degree is produced in samples of whole blood and separated plasma by addition of potassium carbonate. In the example given, this was 15 millimoles per liter of plasma or blood. Each sample of whole blood or plasma is divided into two parts. One sample of whole blood is equilibrated at 37°C with the gas mixture having a P_{CO_2} of 66.0 mm Hg, and the other is equilibrated with the gas mixture having a P_{CO_2} of 28.7 mm Hg. The pH of each sample is measured, and its plasma bicarbonate concentration is calculated. This procedure determines points *1* and *2* on the buffer line of blood having a base excess of 15 millimoles per liter. The points are plotted in figure 54 and a straight line is drawn through them.

The samples of plasma having the same base excess of 15 millimoles per liter are also equilibrated at 37°C with the same gas mixtures. The pH of each sample is measured, and the plasma bicarbonate concentrations are calculated. This gives points *3* and *4* on the buffer line of separated plasma having the known base excess. The points are plotted in figure 54, and a straight line is drawn through them.

The two points intersect at *A*. Since this point is common to the buffer line of plasma of whole blood containing hemoglobin

Table 18

Point	Base Excess mEq/l	HB g%	P_{CO_2} mm Hg	pH	$[HCO_3^-]_p$ mM/l
1	+15	15.2	66.0	7.41	40.5
2	+15	15.2	28.7	7.66	31.3
3	+15	0	66.0	7.37	37.1
4	+15	0	28.7	7.71	35.5
A	+15	15.2 or 0	45.9	7.52	36.4
5	−15	15.2	66.0	7.03	16.9
6	−15	15.2	28.7	7.22	11.4
7	−15	0	66.0	6.87	11.8
8	−15	0	28.7	7.17	10.1
B	−15	15.2 or 0	22.5	7.26	9.8

SOURCE: Adapted from the date of Siggaard-Andersen and Engel, 1960, *Scand. J. Clin. Lab. Invest.* 12:177. (Reproduced by permission.)

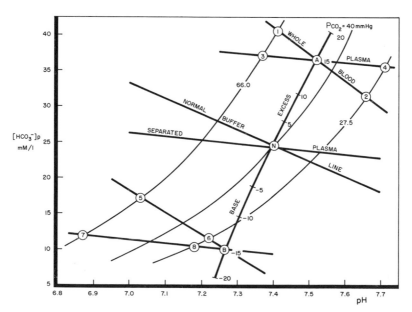

Fig. 54.   Construction of the base excess scale on a pH-bicarbonate diagram. Adapted from the data of Siggaard-Andersen and Engel, 1960, *Scand. J. Clin. Lab. Invest.* 12:177 (Reproduced by permission.)

Table 19

Coordinates of the Base Excess Curve

Base Excess	pH	P_{CO_2}	$[HCO_3^-]_p$
20 mEq/l	7.57	45.2 mm Hg	40.0 mM/l
15	7.52	45.9	36.2
10	7.47	45.1	31.7
5	7.43	43.2	27.7
0	7.38	40.0	23.0
−5	7.34	35.9	18.8
−10	7.30	29.8	14.2
−15	7.26	22.5	9.8
−20	7.24	14.2	5.9

SOURCE: Adapted from the data of Siggaard-Andersen and Engel, 1960, *Scand. J. Clin. Lab. Invest.* 12:177. (Reproduced by permission.)

and to the buffer line of separated plasma, it must be the one point on the graph representing a base excess of 15 millimoles per liter which is independent of the hemoglobin content of the blood. The buffer line of any sample of blood having the same base excess, no matter what its hemoglobin content, must pass through this point. The point then establishes the point representing 15 millimoles base excess on the base excess scale.

The process is repeated with whole blood and separated plasma samples containing other degrees of base excess, and the base excess scale is plotted. The second set of data in table 18 were obtained with blood having a base deficit of −15 millimoles per liter produced by adding hydrochloric acid to whole blood or to separated plasma. The two buffer lines intersect at point B which establishes the −15 millimoles per liter point on the base excess scale. The coordinates of the scale are given in table 19.

The base excess scale can also be plotted on the pH-logP_{CO_2} diagram as shown in figure 55. In that figure the blood buffer lines are plotted from example 19, section 2.4, and from example 25, section 3.2.

The manipulations by which standard bicarbonate and base excess or deficit are determined are performed on blood after it has been withdrawn from the subject. Their determination may assist the physician in identifying and quantitating the acid-base status of his patient, but they cannot substitute for his sound understanding of physiological principles.

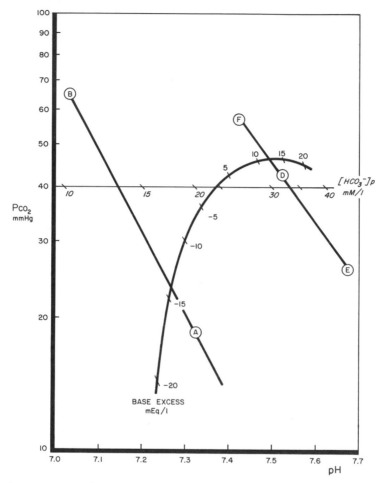

Fig. 55.   The base excess scale plotted on a pH-log P_{CO_2} diagram. The data for the blood of a patient exhibiting metabolic acidosis described in example 25, sec. 3.2, are plotted to show that the base deficit is −14 mEq/l. The data from table 11 obtained on a normal man hyperventilating and breating carbon dioxide after taking sodium bicarbonate are plotted (points D, E, and F) to show that the base excess is 13 mEq/l. Scale adapted from data of Siggaard-Andersen and Engel, 1960, *Scand. J. Clin. Lab. Invest.* 12:177. (Reproduced by permission.)

Bibliography

Brewer, G. J., ed. 1972. *Hemoglobin and red cell structure and function.* New York: Plenum Press.

Brewin, E. G.; Gould, R. P.; Nashat, F. S.; and Neil, E. 1955. An investigation of problems of acid-base equilibrium in hypothermia. *Guy's Hosp. Rep.* 104:177–214.

Comroe, J. H., Jr. 1965. *Physiology of respiration.* Chicago: Year Book Medical Publishers.

Comroe, J. H., Jr.; Forster, R. E., II; DuBois, A. B.; Briscoe, W. A.; and Carlsen, E. 1962. *The lung.* 2d ed., Chicago: Year Book Medical Publishers.

Edsall, J. T. 1972. Blood and hemoglobin: The evolution of knowledge of functional adaptation in a biochemical system. *J. Hist. Biol.* 5:205–257.

Gamble, J. L. 1954. *Chemical anatomy, physiology and pathology of extracellular fluid.* 6th ed., Cambridge: Harvard University Press.

Henderson, L. J. 1928. *Blood.* New Haven: Yale University Press.

Huckabee, W. E. 1961. Henderson vs. Hasselbalch. *Clin. Res.* 9:116–19.

Nahas, G. G., ed. 1966. Current concepts of acid-base measurement. *Ann. N.Y. Acad. Sci.* 133:1–274.

Peters, J. P., and Van Slyke, D. D. 1932. *Quantitative clinical chemistry.* Vol. 1, *Interpretations.* Vol. 2, *Methods.* Baltimore: Williams & Wilkins.

Pitts, R. F. 1968. *Physiology of the kidney and body fluids.* 2d ed. Chicago: Year Book Medical Publishers.

Rispens, P. 1970. *Significance of plasma bicarbonate for the evaluation of H+ homeostasis.* Gronigen: Van Gorcum & Co.

Rossi, L., and Roughton, F. J. W. 1967. The specific influence of carbon dioxide and carbamate compounds on the buffer power and Bohr effects in human hemoglobin solutions. *J. Physiol.* 189:1–29.

Roughton, F. J. W. 1964. "Transport of oxygen and carbon dioxide." Chap. 31 in *Handbook of Physiology, Respiration*, vol. 1, pp. 767–828. Washington, D.C.: American Physiological Society.

———. 1970. *Some recent work with interactions of oxygen, carbon dioxide and haemoglobin*. The seventh Hopkins Memorial Lecture. *Biochem. J.* 117:801–12.

Schwartz, W. B., and Relman, A. S. 1963. Critique of the parameters used in evaluation of acid-base disorders; "Whole-blood buffer base" and "standard bicarbonate" compared with blood pH and plasma bicarbonate concentration. *New Eng. J. Med.* 268:1382–88.

Shock, N. W., and Hastings, A. B. 1935. Characterization and interpretation of displacements of the acid-base balance. *J. Biol. Chem.* 112:239–62.

Siggaard-Andersen, O. 1963. The acid-base status of the blood. *Scand. J. Clin. Lab. Invest.* Vol. 15, suppl. 70.

Singer, R. B., and Hastings, A. B. 1948. Improved clinical method for estimation of disturbances of acid-base balance of human blood. *Medicine* 27:223–42.

Van Slyke, D. D.; Wu, H.; and McLean, F. C. 1923. Factors controlling the electrolyte and water distribution in the blood. *J. Biol. Chem.* 56:765–849.